L—U—X—E
LEADING THE NEW TREND OF MODERN LIVING SPACE

炫 引领现代居住空间新趋势

深圳视界文化传播有限公司 编

中国林业出版社
China Forestry Publishing House

PREFACE
序言

MODERN YET NOT INEXPERIENCED, CONCISE BUT GORGEOUS

现代不显其涩　简约不失其华

Design style is the first attribute which can directly reflect the space label and symbol; space is the main carrier of design style; they supplement each other. As modern style is widely used in designs, more and more people have distinct love for this style. The multiple diverse classifications of modern style are more suitable for modern people's more profound pursuit of living space so that it is more harmonious, more humanistic and closer to daily life. Extending style to residential space design, everyone has a different definition and understanding of style. As a space formation element, modern style reflects a kind of personality.

The presentations of any style are from our common survival experiences. Intuitive aesthetics, man and nature dependent aesthetics and miracle aesthetics in daily life are conveyed through certain special surface texture. In the early 1990s, American "postmodern" designer Michael Graves used "postmodern" design concept in different interior designs including a series of hotel and office building, for example, American Walt Disney theme park. He continued to use his usual "postmodern" technique, removed impurities and retained the essence of design elements and used exaggerated technique to express the essences in mind. For Philippe Starck, the dream creation giant, everyone knows his works; no matter interior or furniture design, every piece of his design is full of humor and wit with unique state furnishings. He is definitely a leader of modern style.

The characteristics of modern style attach great importance to the performance of personality and creation from the perspective of space. It doesn't advocate heaviness but stresses functions. The materials are not limited into a single one; it highlights the structural relationship between materials and strives to present a high-tech interior space atmosphere which is different from traditional style.

Nowadays, no matter residence or hotel, when space becomes the final luxury for human, lifestyle becomes globalizational and the environment becomes complex. Space today is no longer the same as the previous and uses adherent design language to reason; the design process has become more abstract. At that time, modern style will be the only relief of our personal spaces.

When people spend too much time in the crowded city, their inner hearts are yearning to have a space to calm down and deposit mind alone. This kind of tranquil and peaceful living space is everyone's dream while modern concise style fulfills their psychological desires. "Extreme simplicity" or "abundance" only has a slight difference. Design and taste don't need to be attached on fame and wealth. The anima of modern concise space relies on understanding of the designer and the viewer.

设计风格作为最能直接体现空间标签与符号的第一属性，而空间又是设计风格的主要载体，两者相辅相成。随着现代风格在设计中的广泛应用，越来越多的人对这种风格有独特的喜爱。现代风格多元多样性的分类，更能符合当下人们对居住空间更深层次的诉求，从而更为和谐，更为人性，也更贴合日常生活。将风格引伸至居住空间设计，每个人对风格都有不同的定义和理解，而现代风格作为空间构成元素，彰显的是一种个性。

任何风格形态的呈现，无非都源于我们共同的生存体验。对于直观的美感，人和自然依存的美感，日常生活中奇迹发现的美感，然后通过某种特殊的表面材质传达出来。20世纪90年代初，美国的"后现代主义"设计师Michael Graves，将"后现代主义"的设计理念应用于不同的室内设计中，其中包括美国Walt Disney主题公园等一系列酒店及办公大楼。他沿用了他惯常的"后现代主义"手法，将设计元素去芜存菁后，再采用夸大的手法来表达意念中的精华。造梦巨人Philippe Starck，他的作品恐怕无人不知，他的每一项设计不论是室内或家具均充满着幽默与谐趣，装饰中有着独特的形态，是一个彻头彻尾的现代风格引领者。

现代风格的特点从空间上重视个性和创造性的表现，不主张繁重，强调功能性；用材不局限于单一，注重材料之间的结构关系，力求表现出一种完全区别于传统风格的高度技术的室内空间氛围。

现如今，不论住宅或酒店，当空间沦为人类最后的奢侈品，生活方式变得世界化，环境变得复杂，今天的空间将不可如以往那样，沿用固守的设计语言来推理，设计过程已变得更加抽象。那时，现代风格将会是作为我们个人空间的唯一解脱。

当人们太多时间身处于喧嚣的都市，内心都渴望拥有一个能够安静下来，让思想沉淀的空间，一种安静、祥和的居住空间是每个人都追求的目标，而现代简约风格则实现了人们这种心理的愿望。"极度简陋"或"丰裕"只是一线之差，设计及品味是不需要依附在名利之上的，现代简约空间的灵气本来就需要设计师及观者的领悟力。

OneHouse Interior Design / Lei Fang

壹舍设计 / 方磊

CONTENTS
目录

MODERN LUXE 现代奢华

008
COMFORTABLE AND CONSIDERATE MODERN RESIDENCE
舒心解意 现代美宅

016
RETURN TO THE TRUE LIFE
回归生活本真

024
NON INERTIAL MODERN FRENCH RESIDENCE
非惯性现代法式居所

038
METROPOLITAN MODERN FASHION
大都会的摩登时尚

046
MANHATTAN COMPLEX
曼哈顿情结

056
ELEGANT AND LUXURIOUS RESIDENCE
雅奢品居

066
FLOURISHING CITY AND BEAUTY
繁都丽影

076
LIGHT LUXURIOUS FASHION AND GOLDEN DEPOSIT
轻奢时尚 流金积淀

086
GORGEOUS TIME
妍丽时光

094
DREAM TIME ON THE CLOUD
云上的梦幻时光

104
DREAMLIKE PEACH GARDEN, LEISURE ORIENTAL VILLA
梦回桃花园 悠然东方墅

110
HERMES HOME
爱马仕之家

122
ART FROM LIFE
来源生活的艺术

128
POETRY · METAMORPHOSIS
诗意·蜕变

136
LIGHT LUXURIOUS ELEGANCE, LIVING AESTHETICS
轻奢雅致 人居美学

144
FASHION AND CONCISENESS IN ROUND AND SQUARE
于方圆之中 见时尚气质

150
IDEAL ART WITHOUT BOUNDARY
理想的无界艺术

154
WHEN CALMNESS AND GENTLENESS MEET BY CHANCE
当沉稳与温婉不期而遇

MODERN SIMPLICITY 现代简约

164
THE BEAUTY OF BALANCE
平衡之美

178
A CONCISE HOME
简约之家

188
STROLL IN FOREST
丛林漫步

196
TRANQUIL AND WARM SUNSHINE
静谧暖阳

202
FOREST HOME
森林之家

206
DREAM OF ARTISTIC VILLA
艺墅之梦

210
LINEAR ART AND POP ART
线性艺术与波普艺术

216
GREEN MOUNTAIN AND WHITE CLOUD
青山昼·白云栖

224
LIFE WITH LIGHT
过有光生活

232
UNIQUE ORIGINALITY, LEISURE RESIDENCE
独具匠心 悠然居所

236
EVERY FLOOR A REALM AND SIX FLOORS SIX WORLDS
一层一境界 六层六重天

POSTMODERNISM 后现代

250
TIMES WILL
时代的意志

260
SCENORGRAPHY INTERIOR
阁楼·戏剧

272
NOBLE·CLOWN
贵族·小丑

280
TONALITY
格调

288
ONE DAY IN PARIS
巴黎浮生

294
METROPOLIS AND YUPPIE
都市·雅痞

300
EXTREME DEDUCTION
极致演绎

306
BLOOMING SUMMER FLOWERS
夏花似锦

310
FACING THE SEA WITH SPRING BLOSSOMS
面朝大海 春暖花开

MODERN

LUXE

现代奢华

对传统奢华设计概念的一种反思，用低调的方式诠释奢华，注重时尚精致、华美气质，以温馨舒适的居住理念打造全新的奢华感受。

A reflection of traditional luxurious design concept, interpreting luxury in a low-key way, highlighting fashionable exquisiteness and gorgeous temperament, creating a new luxurious feeling by warm and comfortable living concept.

COMFORTABLE AND CONSIDERATE MODERN RESIDENCE

舒心解意 现代美宅

In order to cater to the lifestyle of modern people, the designers specially design a half-open space. Luxurious space needs natural integration; natural changes can make people feel the breath of life. For people who are busy in the city all day long, closing to nature is more yearning. At the same time you can hold a party here. Modern and fashionable decorative style is relaxing and dynamic; having a rest under natural background, home is exactly the place to "act wildly".

Project name: Chengdu Tianhe Zhengxing Villa
Design company: Matrix
Main materials: new moon ancient marble, wood veneer, solid wood floor, leather, wall cloth, stainless steel, mosaic parquet, artistic carpet, gray lens steel, etc.
Chief designers: Guan Wang, Jianhui Liu, Zhaobao Wang
Location: Chengdu, Sichuan
Area: 885m²

项目名称：成都天合正兴别墅
设计公司：矩阵纵横
主要材料：新月亮古大理石、木饰面、实木地板、皮革、墙布、不锈钢、马赛克拼花、艺术地毯、灰镜钢等
主创设计师：王冠、刘建辉、王兆宝
项目地点：四川成都
项目面积：885m²

This project emphasizes the perfect combination of function and form; modernistic design style emphasizes on inner charm of the space and its essence is the emphasis of rationality of functions. The designers attach great importance to the precise, delicate and concise structure, adopt the most advanced technology and keep the original form of natural materials. Stone walls, solid wood floor and the use of leather promote the overall temperament; mosaic parquet and artistic carpet endow the space with more colors, adding fun and romance to the space. Steel and modern furniture configuration bring out the theme of the whole style. The entire design is full of fashionable sense and practical with rich details, which is intriguing.

The tall living room is texture; various materials enrich the space atmosphere; noble stone, modern metal, warm cloth and artistic furnishings make guests feel the identity and good taste of the owner. The bedrooms and study stress privacy and are customized according to the owner's preference. The entire tone of the neutral study is calm; red lines in it are very prominent, which activates the space atmosphere and properly creates an artistic atmosphere. Designs of two bedrooms have obvious order; the master bedroom uses more yellow and metal to manifest the significant manner while the light elegant subaltern room makes people feel the warmth of home.

为了迎合现代人的生活方式，设计师特别设计了半开放式空间。无论多么华贵的空间都少不了自然的融合，自然的变化让人们能够感受到生命的气息。对于整日在都市中忙碌的人们来说，与自然的亲密接触更让人渴求。同时这里也可以来个朋友间的聚会，现代时尚的装饰风格轻松而充满活力，在自然的大背景下尽情地放松，家本就是一个可以让人"撒野"的地方。

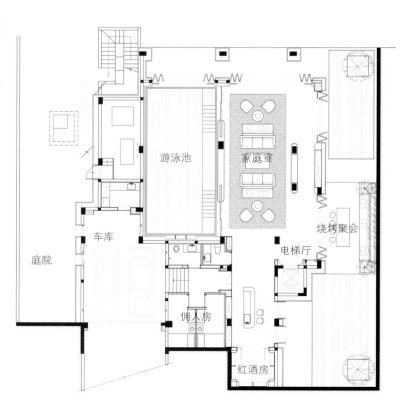

本设计强调功能与形式的完美结合，现代主义设计风格更加强调空间内在的魅力，其本质就是强调功能的合理性，设计师重视结构的精确、细致、简洁，采用了最先进的技术并保持了自然材料的原始形态。石材墙面、实木地板以及皮革的运用让整体气质提升，马赛克拼花，艺术地毯给空间多了一份色彩，增添了空间的乐趣性和浪漫意味。钢材和现代感极强的家具配置让整体风格点题。整个设计兼具时尚品味和实用性，细节丰富，耐人寻味。

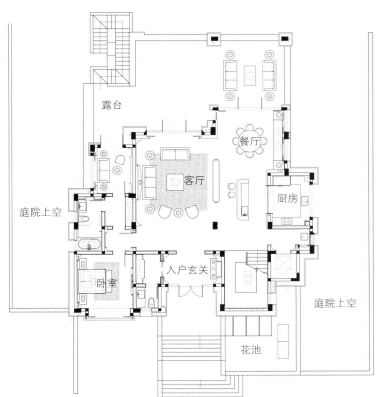

挑高的客厅品质感十足，多种材质的混搭丰富了空间氛围，石材的贵气，金属的现代，布艺的温暖，再装点上极具艺术感的饰品，让访客能够感受到主人的身份和不俗的品味。卧室和书房更强调私密性。这类空间则根据主人的喜好做了定制性的设计。中性的书房整体色调沉稳，点缀其间的红色线条尤为突出，既活跃空间气氛又适度地营造出艺术氛围。两间卧室的设计有很明显的主次之分，主卧用了更多的黄色和金属来彰显不容忽视的气场，而次卧的淡雅则让人感受到家的温柔。

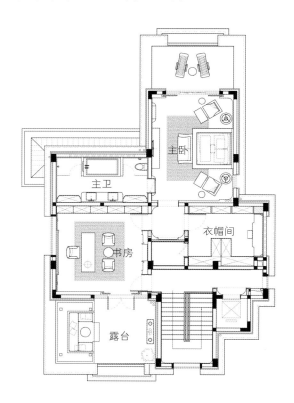

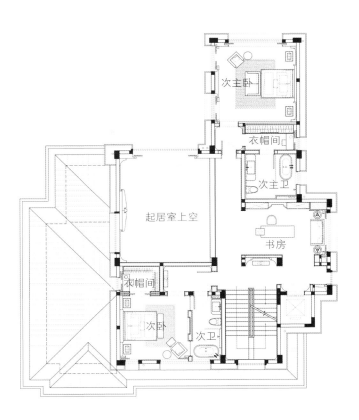

RETURN TO THE TRUE LIFE
回归生活本真

The weeping willow is dancing; the powerful bamboo is fresh and green; flowers are blooming like a piece of brocade; the low thatched cottage relies on the spring pond in the green grass.

Open the door of brushwood, the swallows under the roof seem to be frightened to fly to the pond to draw water; yet the fish don't dodge; it is quite funny.

Here is remote and far away from the hustle and bustle, so the spring is more tranquil; living here, the inner heart is more comfortable and leisure and the warm spring becomes longer.

The birds suddenly stop chirping because of the person's closeness, fly away, then follow the person after a while and keep a right distance, which is interesting.

——Xu Ji

Project name: Suzhou High-Tech Mountain Single-Family Villa
Design company: Xisheng Architectural Design Studio
Main materials: marble, wallpaper, cloth, crystal, etc.
Designers: Yaohua Tang, Wenzhe He, Chenwei Gu
Photographer: Shenfeng Zhu
Location: Suzhou, Jiangsu
Area: 650m²

项目名称：苏州苏高新见山独栋别墅
设计公司：西盛建筑设计（上海）事务所
主要材料：大理石、壁纸、布艺、水晶等
设计师：唐瑶华、何文哲、顾陈玮
摄影师：朱沈锋
项目地点：江苏苏州
项目面积：650m²

This *Mountain Abode* is a poem with five characters to a line written by the famous poet Xu Ji in Southern Song Dynasty. It depicts a perfect integration of natural environment and house and expresses the rustic charm of the indifferent owner who lives in a remote place and the enjoyment of man and nature.

Through the ages, only landscape is none the worse for being twice seen. Living near mountains is not only the lifestyle loved by the ancient Chinese, but also the nature and comfort brought by modern people's far away from the madding crowd and enjoyment of living a mountain life. Go out, you can see all the prosperity; go in, you can enjoy the tranquility alone. This is perhaps the moods and feelings which modern people are lack of.

Do you who have a successful career in the city want to free your mind and put your home in the quiet and tranquil landscape?
The designers of this project use unique humanistic feelings to create this single-family villa. Living near green mountains and water is full of texture and fashionable taste; even the air is full of the temperament of the era.

柳竹藏花坞，茅茨接草池。
开门惊燕子，汲水得鱼儿。
地僻春犹静，人闲日更迟。
小禽啼忽住，飞起又相随。
——徐玑

这是一首由南宋著名诗人徐玑所创作的五言绝句《山居》。诗中描绘了自然环境与屋舍完美融合，表达了作者心远地偏的野趣以及人与自然相得其乐的欢愉。

古往今来，唯有山水不厌看。见山而居，不仅是中国古人钟情的生活状态，也是现代人远离尘嚣，享受山居生活带来的自然与舒适。出则繁华尽揽，入则静谧独享。这也许就是现代人居所缺少的心境与情怀。都市里拥有成功事业的你，是不是也想放飞心灵，把家安在宁静致远的山水之间呢？

本案设计师以独具特色的人文情怀，打造出这套独栋别墅。依青山绿水而居，极富质感与时尚品味，空气中弥漫着时代的气质。

NON INERTIAL MODERN FRENCH RESIDENCE
非惯性现代法式居所

French thought and history had a decisive shaping effect on values of other countries. In Louis XIV period, Versailles was the incomparable aesthetics and political model of European court; the revolution epic in late 18th century and early 19th century inspired national liberators all over the world; in 19th century, *The Napoleonic Code* was adopted by the newly independent country widely. Until today, the France morphology originated in the 14th century is still widely referred and continued in world art after several hundreds of years of evolution. The popularity of French architectural style in China is one of the best example. Shanghai Xuhui Western Suburb Villa continues French spirits and aesthetics for hundreds of years with absolute French architecture and hard decoration style.

Project name: Shanghai Xuhui Western Suburb Villa
Design company: LSDCASA
Design team: LSDCASA Business Unit One
Location: Shanghai
Area: 480m²

项目名称：上海旭辉西郊别墅
设计公司：LSDCASA
设计团队：LSDCASA事业一部
项目地点：上海
项目面积：480m²

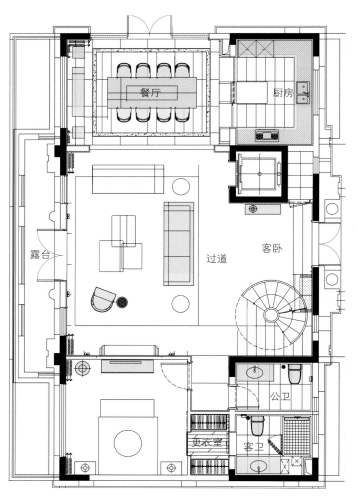
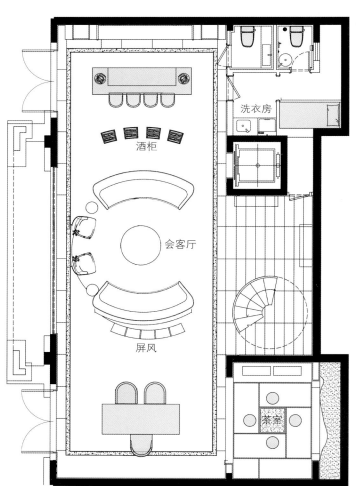

Facing with heavy French civilization, LSDCASA chooses modern concise lines to break the limitation of style, create a non inertial French residence and make a new birth of modern French style. Easily handling difficulties is not flightiness but the confidence with high skills.

In the entire space design, the diluent exquisiteness increases the profound connotation and shifts skillfully between more and less, deep and shallow and classical and innovation. The furniture materials and patterns abandon complicated designs; neutral tones balance infinite comfort in proportion, emotion and story. What used throughout the space are pieces of gray blue and jumping gold lines. Gray blue comes from the Seine which breeds Paris culture while gold comes from the gold edges of sunflowers outlined by Vincent van Gogh.

The DNA of designs in the living room is "romance", "elegance" and "delicacy", which are essences of French culture and aesthetics. Gray and white voice-print carpet paves the texture of the space. The couch chooses FENDI brand with soft tone and material, collocating with gold and bright tea table, which is harmonious. The modeling and grain of the dining chairs are originated from the formal dress of French banquet in 17th century; the unique light purple and long dining table with light gloss forms an elegant first impression. The gray blue curtain with distant view and gold sideboard cabinet form a bright color ornament in the space. The master bedroom is a fresh abode for gentleman.

Without flatulent decorations and colors, all attitudes about tastes are communicated in the tranquil and moderate black, gray and gold tones. As a furnishing, the carpet even has less sense of decoration of curves. Soft curtains, hard decoration elements and corners and grain details of furniture can trace back to brilliances of French romance.

For elder parents, the design uses plain and elegant tones and graceful texture; in some non-visual main parts, it boldly uses black gold line furniture and carpet with obvious contrast, which adds vivid fun. Design of the boy's room starts from the concept of "small gentleman"; black, white and gray tones basically stabilize the visual tonality. Design of girl's room doesn't stress "child interest", but adds artistic atmosphere; large amount of macarons and pure and transparent crystal furnishings add a sweet romance of French dessert.

The reception room in the negative layer is the space for the owner to self-release. The crystal droplight with a diameter of only 2.8 meters and two big arc sofas are light elegant, exquisite, jumping and harmonious. Tones of curtain and carpet echo distantly. Design of the bar spans from French style to golden age in New York. Classic gentleman colors such as black, white, gold and coffee exist in harmonious proportion. The entire space visual tonality is as dark as the intoxicating night. Book bar in the reception room uses various collections to highlight elegant emotional appeal; the black and white geometric patterns of the carpet circulate as if never-ending happiness and inspiration.

法国的思想和历史曾对其他国家的价值观产生了决定性的塑造作用，路易十四时期，凡尔赛宫是欧洲法院无与伦比的美学和政治典范；18世纪末和19世纪初的革命史诗，启发了全球的民族解放者；19世纪，《拿破仑法典》被新独立国家广泛采纳。直到今天，源自于14世纪的法兰西形貌，经由数百年演变，仍在世界艺术中被广泛提及和延续。法式建筑风格在中国的流行就是一个最好的例证。比如上海旭辉西郊别墅，承袭数百年的法式精神和审美，绝对的法式建筑和硬装风格。

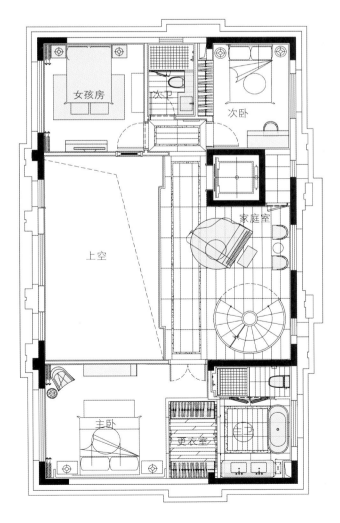

  LSDCASA 面对厚重的法式文明，选择以现代的简明线条勾勒，打破风格的禁锢，打造非惯性法式居所，让现代法式新生。举重若轻，却非轻薄，而是四两拨千斤的自信。

  整体空间设计中，以冲淡的精致叠加深厚的底蕴，在多与少，深与浅，古典和创新之间，轻柔踱步。在家具材质和款式方面摒弃繁复浮夸，中性的色调在比例、情绪和故事间平衡出了无限的舒适。贯穿其中的，是成块的灰蓝和跳跃的金色线条。灰蓝色来自孕育巴黎文化的塞纳河，金色来自梵高勾勒向日葵的金边。

  客厅的设计 DNA 是"浪漫"、"优雅"、"精致"，正是法式文化和美学的精髓。灰与白的声波纹地毯铺垫了空间的质感。长沙发选用 FENDI 品牌，色调与材质均偏柔和，搭配以金属亮色的茶几，形成协调。餐椅的造型与纹路，源自 17 世纪法式宴会的礼服，独特的浅紫色与散发淡淡光泽的长餐桌，形成优雅的第一观感。远景的灰蓝色窗帘和金属色的餐边柜，对空间形成亮色点缀。主卧是鲜明的绅士之居。没有浮夸的装饰或色彩，一切关于品味的态度均在平静、温和的黑、灰、金色调中被沟通。

除了作为点缀的地毯，甚至少有曲线的装饰感。在柔和的窗帘、硬装元素和家具的一些边角、纹理细节，能追溯到法式浪漫的点滴光辉。

对年长的父母，在设计中以素雅的色调和雅致的质感点缀平和，但在一些非视觉主体的部分，大胆运用了黑金线条家具及对比感明显的地毯，添加生动的趣味。男孩房的设计从"小绅士"的概念入手，黑白灰作为主色基本稳定了视觉调性；女孩房的设计不注重"童趣"，而是更多地添加了艺术氛围的营造，马卡龙色的大量运用，水晶饰品的纯净通透，添加一丝法式甜品的甜腻浪漫。

在负一层的会客室，是主人释放自己的空间。2.8米直径的水晶大吊灯，和两个弧形大沙发，淡雅中精致，跳脱中和谐。窗帘和地毯的色调遥相呼应，酒吧的设计从法式跨越到纽约的黄金时代。黑、白、金、咖的经典绅士色彩，以和谐的比例存在。空间整体视觉调性偏暗，如同醉人的夜色。会客室的书吧以各种藏品突出文雅的情调，地毯的黑白几何图纹循环往复，如同永不停歇的快乐与灵感。

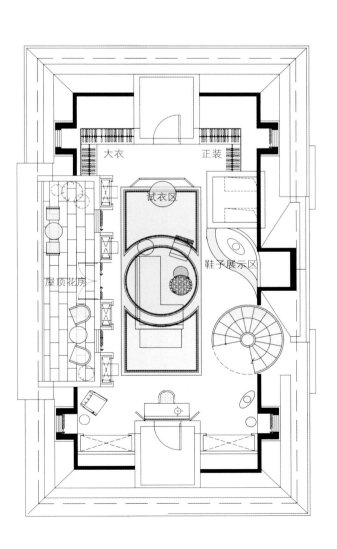

METROPOLITAN MODERN FASHION

大都会的摩登时尚

MODERN SIMPLICITY ■ 现代简约

Swan Lake is a landmark high-end residential project in Chinese Town, Shunde, Foshan. Located in the beautiful scenery line of the city center, it has 8000 acres mountain scenery, with Wetland Park in the south and Guipan Lake and Qingyun Lake in the west. Eric Tai design team upholds low-key and luxurious design concept, integrates humanistic and natural elements, elegantly expresses and sublimates metropolitan modern fashion and diverse style temperament and creates a spiritual elegant enjoyment with exquisite detail performance. High-quality furniture matches with special fashion aesthetics, adding beauty with outside beautiful landscape.

Project name: Foshan Chinese Town Swan Lake Show Flat
Interior design: Eric Tai Interior Design
Designer: Eric Tai
Photographer: Yanming Chen
Location: Foshan, Guangdong
Area: 250m²

Main materials: white sand beige marble, Canadian maple wood veneer, peacock blue jade marble, gold mounted jade marble, green bronze stainless steel, walnut wood veneer, wallpaper, oak wood floor, etc.

项目名称：佛山华侨城天鹅湖样板房
室内设计：戴勇室内设计师事务所
设计师：戴勇
项目摄影：陈彦铭
项目地点：广东佛山
项目面积：250m²

主要材料：白沙米黄云石、加拿大枫木木饰面、孔雀蓝玉云石、金镶玉云石、青古铜不锈钢、胡桃木饰面、墙纸、橡木木地板等

The space design is eclectic with fluent and neat techniques. Warm earth color stresses the warm sense of belonging; dark brown injects calm tone. The living room is equipped with sofas and chairs which have suitable sizes according to the space dimension to ensure complete functions and make the space neat and clear. The dining room chooses round Chinese table and wood veneer dining chairs; the nearby fashion bar matches with oak wood decorative shelf with exquisite wine bottles and cups, creating a relaxing and elegant wine tasting space. The living room, dining room and bedrooms are equipped with intelligent household system, providing modern and convenient lifestyle for the owner.

Green bronze metal lines are used throughout the home, which enriches material details of the space along with special customized art glass and warm wood. As for details, Hermes cashmere blanket, tea sets and sculpture furnishing articles by artist Qu Guangci collocate perfectly. The designer simplifies complicated materials, thinks about the nature of space design and creates a concise and neat honorable space.

天鹅湖是佛山顺德华侨城的标杆性高端住宅项目，坐落在城市中央美丽的中心风景线上，拥揽530多公顷顺峰大境，南临湿地公园，西临桂畔海、青云湖。戴勇设计团队秉承着低调奢华的设计理念，融入人文与自然元素，优雅地表达和升华了大都会摩登时尚和多元融合的风格气质，同时加以精致的细节演绎，成就一种充盈于心灵之中的优雅享受，高品质的家具搭配呈现矜贵非凡的时尚美学，与室外湖光山色美景交相辉映。

　　空间的设计不拘一格,手法流畅利落。以温和的大地色系强调温馨的归属感,深棕色彩注入沉稳的格调。客厅中选择体量与空间适应的沙发、座椅等家具保证使用功能完善的同时,也让空间显得游刃有余。餐厅选用圆形中餐桌及木皮材质结合的餐椅,毗连的时尚吧台配衬橡木装饰层架上精致的酒瓶与酒杯,缔造轻松雅致的品酒空间。客餐厅和卧室均设有智能家居系统,为户主提供现代化且恬然便捷的生活方式。

　　青古铜色金属线条贯穿居所的各处,与特殊定制的艺术玻璃、温润的木材丰富了空间的材质细节。细节之处,爱马仕的羊绒毯、茶具与艺术家瞿广慈的雕塑摆件悉心配搭。设计师删繁就简,思考空间设计的本质,创造出简约利落的尊享空间。

MODERN LUXE ■ 现代奢华

MANHATTAN COMPLEX
曼哈顿情结

Though Shanghai and New York are separated by the whole Pacific Ocean, they are like a pair of twin cities, lead the economy trend of the world nowadays and promote the development of global cities. The Bund and Manhattan are the souls of the two cities. They are the same prosperous and modern, constantly integrate the most vanward life concept in the world and play important roles in the development of human city. This time the designer Lei Fong cooperates with China Resources Property to design this show flat and bring Manhattan metropolitan style into Shanghai. This kind of unique design and exquisite works of art make the space present modern and avant-garde tone and full of elegant temperament like the art museum.

Project name: The Bound of Bund Show Flat
Design company: OneHouse Interior Design
Main materials: stone, upholstery, leather, bronze metal, wallpaper, etc.
Designer: Lei Fang
Cooperating designer: Yonggang Ma
Location: Shanghai
Area: 450 m²

项目名称：华润外滩九里样板房
设计公司：壹舍设计(方磊设计师事务所)
主要材料：石材、扣布、皮革、古铜金属、墙纸等
设计师：方磊
参与设计：马永刚
项目地点：上海
项目面积：450 m²

Urban elites are a group of people who pursue material lives and have artistic tastes. To design spaces for them is a challenge to designers' understanding of art. The designers deeply know that what urban elites pursue is not only a luxurious house but also a perfect combination of material life and artistic taste.

Design of the living room is eclectic and the technique is bold and free. The designers inject dazzling orange and noble purple into the space and use modern splicing carpet to collocate with classical art sculpture, which is like Manhattan's broad feelings of collecting global multi-cultures and containing different styles in different countries, presenting dreamlike colors and extraordinary elegant temperament of international metropolis. Modern geometric gold round droplight in the dining room matches with emerald dazzling tableware, making the space shine luxurious and modern intoxicating magic of the international metropolis. The dining room and kitchen are linked by the western kitchen bar; the bar, family hall, dining room and Chinese kitchen form a strong interactive relations with a panoramic view of the Bund scenery. The freestanding elevator lobby displays works of art from Simard Bilodeau art gallery, making the space bloom as the art museum.

The spacious and bright master bedroom ensures the living comfort with sense of space. Background wall and top wall are linked naturally by arc designs. Gilding blue velvet pillows and art paintings on the wall bring residents luxurious feelings of texture residence and metropolitan plots of converging world tide front-end culture. Silent white and quiet blue in the study match with paintings full of Oriental flavors, creating a warm and comfortable reading and leisure environment. Bathroom of the master bedroom has an international sense and mixes Shanghai style. Bathing area design breaks the original pattern; the designers use large piece of French window to bring sunshine and beautiful scenery into interior, giving the heart the baptism of the nature. The ground mirror in the cloakroom extends the space and brings outdoor scenery into interior. Large area of circular terrace links up every space. The terrace with dining area and several recreational areas can completely satisfy various parties of celebrities and elites. Here you can appreciate the beautiful scenery of the Bund, forget the intense urban life and enjoy the comfort and leisure.

　　上海与纽约虽然隔着整整一个太平洋,却像一对孪生城市一样,引领着当今世界的经济风潮,推动着全球城市的发展。外滩和曼哈顿是两座城市的灵魂,他们一样的繁华、摩登,不断地融合全球最先锋的生活理念,成为人类城市发展史上极其重要的角色。这次,设计师方磊与华润地产合作,设计华润外滩九里的顶平样板房,将曼哈顿的大都会风带入上海。这样的独特设计加上精致的艺术品陈列,使得这个空间既展现出现代前卫的格调,又充满了艺术博物馆般的优雅气质。

　　都市精英是既要求物质生活,又有一定艺术品位的群体。为他们做空间设计,非常考验设计师的艺术领悟力。设计师深知都市精英们所追求的居所不单是一座豪华的房子,更是要将物质生活和艺术品味做完美的结合。

　　客厅空间的设计不拘一格,手法大胆自由。设计师将耀眼的橘色、高贵的紫色融入空间中,用现代感十足的拼接地毯搭配古典韵味的艺术雕塑,宛若曼哈顿汇集全球多元文化包容各国不同风情的宽阔情怀,展现出国际都会的梦幻色彩与超脱浮华的高雅气质。餐厅现代几何造型的圆形金色吊灯搭配色彩如祖母绿般耀眼的餐具配饰,让这个餐厅空间闪耀着国际都会奢华摩登的醉人魔力。餐厅和厨房由西厨吧台连接,吧台与家庭厅、餐厅及中厨形成强有力的互动关系,将外滩美景尽收眼底。独立式的电梯前厅陈列着西马德－比洛多画廊的艺术品,让空间像艺术博物馆一样绽放。

　　宽敞明亮的主卧空间感十足,保证居住的舒适性。墙背景与顶面用弧线设计自然衔接。烫金花纹的蓝色天鹅绒抱枕与墙上艺术画,无不时刻给居住者带来品质居所的奢华感和汇聚世界潮流前端文化的都会情节。书房肃静的白色与一抹幽蓝搭配东方韵味的挂画营造出温馨舒适的阅读休闲环境。主卧卫生间具有国际感又融合了海派风格。沐浴区的设计打破了原始格局,设计师采用大片的落地窗形式将阳光与美景引入室内,让心灵得到自然的洗礼。衣帽间落地明镜将空间延伸并把户外美景带入室内。

　　大面积的环形露台衔接每个空间。露台分布着餐区以及若干休闲区域,完全可以满足名流和精英们的各种聚会。在这里欣赏着上海滩的美丽景色,似乎可以忘记紧张的都市生活,只享受一份惬意与悠然。

ELEGANT AND LUXURIOUS RESIDENCE

雅奢品居

This project is located in Beijing where numerous social elite and middle class people dreamed to live. They have good cultural connotations, cultivations and they pay attention to life tone and have a good taste. That is the reason why the designer sets the theme to be modernized elegant luxurious style. The concise and fashionable decorations, metal color and sense of lines create extreme visual effects and tensioned space feelings. The more magnificent and fashionable design method presents luxurious flavors of multicultures and diverse lifestyles of modern Hong Kong style.

Project name: Beijing Metroplis Jiayuan A2 Show Flat
Design company: EHE DESIGN
Main materials: walnut, burl, stainless steel, marble leather, flannel cloth, etc.
Designer: Ma Jingjin, Xu Yanwen, Ge Xulian, Ma Hui
Photographer: A Guang
Location: Beijing
Area: 316m²

项目名称：北京名都嘉园A2样板房
设计公司：杭州易和室内设计有限公司
主要材料：胡桃木、树瘤、不锈钢、大理石皮革、绒布布艺等
设计师：麻景进、许彦文、葛旭莲、马辉
摄影师：阿光
项目地点：北京
项目面积：316m²

When we look into the interior we could see quite a few of new architectural materials such as stones and metals, which applied around the living room to create luxurious temperament. Monochromatic color such as black, white and gray integrate with concise metal texture lines to build a noble and magnificent atmosphere, which looks fashionable and avant-garde.On the other hand, the choice of furniture is also very fastidious, started from materials to patterns, to make sure every single detail is appropriate. The quality of materials used in this space stand for luxurious and extravagant lifestyles. In addition, several fashionable and charming furnishing and display elements are loved by fashion elites.

In the elegant and luxurious kitchen, whether it is a knife or a wine cup, it is carefully picked by the designer. Open and bright western kitchen designs make life more comfortable. Many exquisite and delicate life details reflect luxurious life movement.

Opening the door of the master bedroom, the sedate and sober gray tone is as if the restrained and elegant temperament of the owner, which emits unique charms. The grayish purple ornament at the bedside is as if the charming smile of the hostess. The fascinating charms are pervaded everywhere.

In the subaltern room, fashionable and noble champagne gold matches with exquisite gray beddings, encountering with modern and fashionable decorative paintings, which creates low-key, elegant and luxurious life.

Entering the underground garage, fashionable and dynamic locomotive model and collections are distinct. Life itself is interesting. These unusual hobbies reflect the owner's pursuit of life.

该项目位于皇城根下，无数社会精英及中产阶级都曾梦想生活在这里，他们拥有良好的文化底蕴和修养，特别注重生活格调及品味。也正如此，设计师将设计的主题定为现代雅奢，它以简洁而不失时尚的装饰，运用金属色和线条感营造出极致的视觉效果和极具张力的空间感受，以更大气、更前卫的设计手法展现出多元文化和多样生活方式带来的现代港式奢华情调。

放眼望去，客厅里大量的新金属、新建筑材料、石材等材质相映成辉，共构奢华气质。黑、白、灰的色彩基调，融合各种金属质感的简化线条，高贵而大气，时尚而前卫。家具的选择也十分考究，从材质到款式的每一个细节都做到恰到好处，细腻缜密的布艺、木材及金属的优秀品质，是豪华、奢侈生活方式的象征。还有各种带有时尚魅力的陈设元素，极受追求时尚的精英人士的喜爱。

高雅奢华的厨房中，无论是一把刀，还是一个酒杯都经过设计师的精挑细选。开阔明亮的西厨设计让生活更加惬意，多处精巧细致的生活细节交汇出华贵的生活乐章。

推开主卧的房门,沉稳冷静的灰色调一如主人内敛而优雅的气质,散发出独特的魅力。而床头点缀的那抹灰紫调就像是女主人迷人的微笑,风情万种就在这一点一滴中弥漫。

次卧中时尚贵气的香槟金色与材质讲究的灰色调床品精彩混搭,巧妙邂逅现代时尚的装饰画,打造低调的雅奢生活。

走进地下车库,发现时尚动感的机车模型和用品收藏,个性范儿十足。生活本身就是丰富有趣的,这些与众不同的爱好正是主人对生活极致追求的体现。

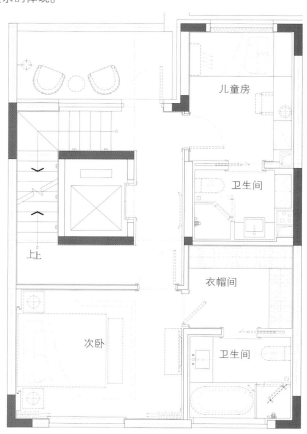

FLOURISHING CITY AND BEAUTY

繁都丽影

It upholds the concept of creating a high-quality residential lake villa and defines life meaning of Jinhua, "Small West Lake". II three interior design which has never been limited by traditional style finds a new balance between art and luxury in innovation and integration. How to amplify core value of life through soft decoration furnishings and bring new meanings to a traditional luxurious design is the key point of II three interior design. Through reshaping color relations of life and reconsidering materials of interior hard decoration and various design elements, the designers explore an aesthetic concept different from traditional luxurious style and collide a distinct design spark.

Project name: Jinhua Baoji Lake Seawall Manor
Design company: II three interior design Co., Ltd
Item designers: Hui Chen, Ang Gao, Miaoqing Xu
Main materials: burl, titanium gold bar, gray marble, etc.

Hard decoration designer: Lei Ying
Soft decoration designer: Licong Zhang
Photographer: Taotao Wan

Location: Jinhua, Zhejiang
Area: 450m²

项目名称：金华保集湖海塘庄园
设计公司：金华市贰三室内设计有限公司
分项设计师：陈卉、高昂、徐苗青

硬装设计师：应磊
软装设计师：张丽聪
摄影师：万涛涛

项目地点：浙江金华
项目面积：450m²
主要材料：影木、钛金条、灰色大理石等

秉承打造湖滨别墅的高品质住宅理念，定义金华"小西湖"的生活意义，从不被传统风格说束缚的贰三设计，在创新与融合中，找到了一种艺术与奢侈的新平衡。如何通过软装的陈设放大生活的核心价值，同时为一种传统的奢侈风格带来设计的新意，是贰三设计思考的重点。通过重构生活的色彩关系，重新考量室内硬装的材质、多样的设计元素，探索出不同于传统奢侈风格的美学概念，碰撞出了不一样的设计火花。

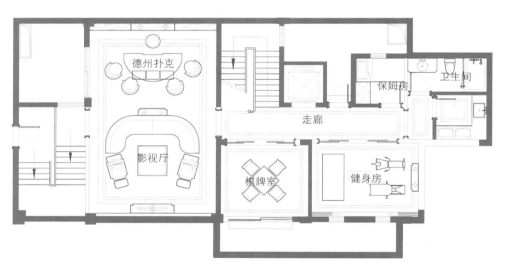

LIGHT LUXURIOUS FASHION AND GOLDEN DEPOSIT

轻奢时尚 流金积淀

Light is a kind of elegant attitude, low-key and comfortable, noble and elegant; luxury is an attitude, a state of pursuing quality and exquisite life. Modern light luxurious life is a fashionable attitude and pursues high quality life enjoyment. It represents a kind of life concept and life attitude, a most perfect interpretation of future life and art. This is the redefinition of light luxurious life in the new era background.

Based on light luxurious culture, space display explores fashionable and modern life state and uses carefully customized furniture and art collections to convey urban upstarts' pursuit of high quality life and attitude towards them. Elegant and luxurious space temperament and new art pioneer reach a balance and integrate, creating an elegant tone for the space.

Project name: Shanghai Vanke Emerald Riverside Project
Design company: Purple's Design Interiors
Main materials: marble, stainless steel, wallpaper, wood veneer, bright copper, etc.
Design team: Purple's team
Photographer: Zhigang Liang
Location: Shanghai
Area: 300m²

项目名称：上海万科翡翠滨江项目
设计公司：北京紫香舸装饰设计有限公司
主要材料：大理石、不锈钢、壁纸、木饰面、亮铜等
设计师：紫香舸团队
摄影师：梁志刚
项目地点：上海
项目面积：300m²

The space colors are implicit and elegant with tranquil and elegant beige as the main tone, collocating with gorgeous and splendid gold, noble and mysterious purple and deep and calm black; the neutral tone balances an appropriate comfortable sense. The furniture abandons complicated and flatulent modeling, concise and decent, stressing choices of crafts and materials. At the same time, many elegant and funny furnishings are placed skillfully in the space; fancy bedside lamp holder, abstract paintings, novel crystal sculpture and various art collections form a rich aesthetic charm with visual art.

To maximize the space, the designer integrates the living room, dining room and study into a whole. In the living room, gorgeous chandelier and furnishings, luxurious and graceful. Decent and comfortable cloth sofa, simple and smooth tea table, fashionable leopard armchair and metal odd chair integrate skillfully, forming the temperament skeleton of the space. Copper which emits charming metal sense promotes the temperament of the space; peacock furnishing articles is graceful, gorgeous and dazzling. In the dining room, gold droplight is luxurious and gorgeous, echoing with the droplight in the living room and manifesting a magnificent and luxurious noble demeanor. Chic candle holder and exquisite tableware present the organic integration of art and life and deliver the temperament and splendidness of the space. the study has art collections from all over the world, at the same time these works of art become parts of the interior decorations, overlapping an artistic tone. Contemporary art is integrated into painting, pendulum clock, candlestick, stean and metal furnishings, which become important visual symbols of the space temperament and create a space atmosphere which is close to life and obviously beyond life.

079

轻，是一种优雅态度，低调、舒适，却无损高贵与雅致；奢，是一种态度，一种追求品质和精致生活的状态。现代轻奢生活，是拥有高雅的时尚态度，并不断追求高品质的生活享受。它代表的是一种生活理念和生活态度，是对未来生活与艺术融合最完美的诠释。——这是新时代背景下对轻奢生活的重新定义。

空间陈设站在轻奢主义文化之上，对时尚现代的生活形态进行探索，运用精心订制的家具和艺术感强烈的收藏品，传达都市新贵对未来高品质生活的追求和态度。优雅奢适的空间气质与新锐艺术的先锋感平衡融汇，创建了该居所的高雅格调。

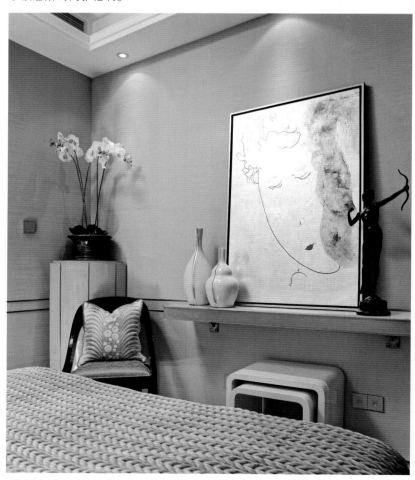

　　空间色彩含蓄优雅，以恬淡雅致的米色为色彩基调，搭配华丽绚烂的金、高贵神秘的紫、深邃冷静的黑，中性的色调平衡出恰当好处的舒适感。家具款式方面摒弃繁复浮夸，简洁大方，注重工艺及材料的选择。同时，各种风雅有趣的摆件在空间中巧妙地布置着：别出心裁的床头灯座、抽象的挂画、新颖的水晶雕塑、新奇多样的艺术收藏品，构成了视觉艺术无比丰富的审美魅力。

　　为实现空间的最大化，设计师将起居室、餐厅、书房合而为一。起居室内，流金溢彩的水晶灯与装饰，沉醉奢华而又雍容尔雅。大方舒适的布艺沙发、简单的光面茶几和时尚的豹纹单人沙发、金属单椅巧妙地结合，构成空间气质骨架。散发迷人金属感的铜提升空间气质，孔雀摆件姿态怡然，绚丽夺目。餐厅中，金色吊灯华丽璀璨，与起居室内的吊灯相呼应，展露出瑰丽奢华的新贵风范。别致的蜡烛灯座及精致的餐具，呈现艺术与生活的有机融合，传递出空间的气质和精彩。而书房里不仅放置了来自世界各地的艺术收藏品，同时这些艺术品也成为室内装饰的一部分，交叠出艺术的格调。当代艺术被融入到挂画、摆钟、烛台、陶罐、金属摆件等细节，分化为空间气质的重要视觉符号，形成接近生活又明显高于生活的空间氛围。

GORGEOUS TIME
妍丽时光

Splendid Yangtze River is a key project of Wuhan Shimao. In order to highlight the reasonable functional layout and space size of the house type, the designers make big adjustment on the plane optimization phase, enlarge the kitchen area and combine western kitchen with dining room, which is convenient for dining and makes the space atmosphere more current and transparent. The ceiling and floor of the living room don't have overmuch decorations and overlapping in order to enjoy the superior scenery outside the living room on the river sides when entering the house. The gold V-shape decoration of the background, rhombus cabinet handle and square droplight in the bedroom bring fashionable and modern sense. The designers use metal and mirror on the basis of the traditional stone and wood veneer, which increases texture of the space with a strong visual impact. The whole space is magnificent, sedate and dynamic.

Project name: Wuhan Shiman Splendid Yangtze River
Design company: PFD+
Photographer: Wenjie Hu
Main materials: stone, tile, metal, mirror, leather, cloth, etc.

Chief designer: Patrick Fong
Cooperating designers: Mengli Zhen, Fu Shi

Location: Wuhan, Hubei
Area: 290m²

项目名称：武汉世茂锦绣长江
设计公司：方振华创意设计（杭州）有限公司（PFD+）
摄影师：胡文杰
主要材料：石材、瓷砖、金属、镜面、皮革、布艺等

主持设计师：方振华
参与设计师：郑蒙丽、施甫

项目地点：湖北武汉
项目面积：290m²

锦绣长江是武汉世茂的重点项目之一。为了凸显其户型功能布局合理和空间大小，设计师在平面优化阶段做了大的调整，扩大厨房面积，将西厨与餐厅结合。不仅方便就餐，同时空间氛围更加流通和通透。客厅的吊顶和地面没有过多的装饰和叠级，是为了入户就可以欣赏到客厅外的一线江景，饱览两岸风情。卧室背景墙的金色V形装饰、菱形柜门把手、方形吊顶带来时尚感的同时，现代感丝丝入扣。设计师在传统的石材和木饰面基础上增加了部分金属以及镜面，增加空间材质丰富感，视觉冲击力十足。整体空间大气、稳重，又具有活力。

DREAM TIME ON THE CLOUD
云上的梦幻时光

"Jiangnan is a place with many beautiful girls while Jinling is a place where many emperors lived." As ancient capital of six dynasties, Nanjing has a history of city for 2600 years and a history of capital for nearly 500 years, stresses culture and education, is known as "cultural hub of the world" and is referred as "two world classical civilization centers" with Rome, which attracts many refined scholars and forms a unique culture of the southern dynasty. Architecture is dead while design is alive. Kenneth Ko always advocates designs with temperature. He rationally grasps the history of Nanjing, deconstructs Chinese and Western cultures and integrates them into show flat design. He uses metropolitan style full of post-industrial avant-garde sense and perceptual humanistic colors, boldly expresses the flouring trend and interprets fashionable and modern life concept.

Project name: Nanjing Minmetals Yanshan Show Flat
Design company: KKD
Main materials: marble, solid wood, special glass, wood veneer, cloth curtain, etc.
Designer: Kenneth Ko
Photographer: A-Kuei
Location: Nanjing, Jiangsu
Area: 240m²

项目名称：南京五矿晏山居样板房
设计公司：深圳高文安设计有限公司
主要材料：云石、实木、特种玻璃、木饰面、布帘等
设计师：高文安
摄影师：阿贵
项目地点：江苏南京
项目面积：240m²

Designs of the recreation hall in the basement are endowed with heavy artistic flavors and space humors which leave an impression on people. The classical painting which presents Parthenon temple relic and lives of Western European gentlemen and ladies is designed by Italian artist. Avant-garde cloud decorations and bronze animal heads full of medieval European amorous feelings are integrated with fashion furniture in Chinese red, which manifests Chinese and Western cultural deposits and interprets concise, luxurious and elegant style and connotation of modern metropolis.

"Mist envelops the cold water, and moonlight the sand; not far from a tavern I moor my boat in the Qinhuai at night." The Qinhuai River flows romance and elegance of ancient Nanjing city for one thousand years. Designs of bar in the negative floor are inspired by the beautiful Qinhuai night; the jumping color is like the dancing thirteen beauties; Chinese pottery and Easter island giant head metal chair combine Oriental and Western elements and depict enchanting charm and sexy of metropolis.

Nanjing was called stone city whose beautiful time is in autumn. "When in autumn rain parasol leaves were shed", delicacy and grace of girls here are presented to the extreme. The designer integrates stone town autumn rain elements into the corridor design; tall and low crystal glass balls pour down like autumn rain, deducing a touching rhyme that rain and wind in autumn make residents intoxicated.

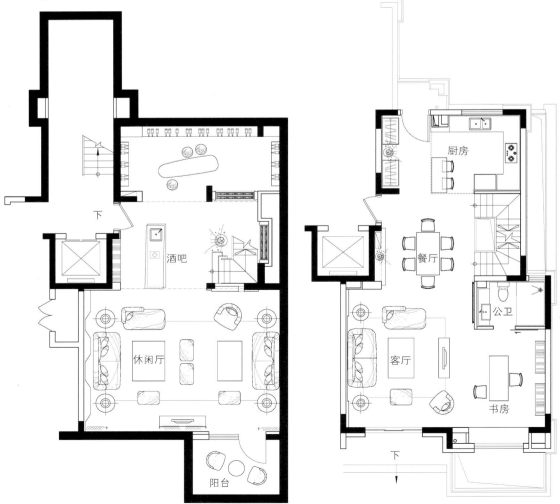

The poet Li Bai described Nanjing as "the topography is suitable for emperor to live and the mountain is located in a strategic place". The Purple Mountain scenery is the most exquisite landscape. Life space designs in the first floor use Purple Mountain as image. Open living room and dining room link together with perfect sights and sceneries. As for furniture colors, the red and purple living room presents nobility. The dining room is full of the beauty of visual conflict and is the best representative of modern people's personal and casual lifestyles.

There were many gifted youths since ancient times who liked drawing landscape ink painting. Kenneth Ko takes lessons from the ancients and injects Xuanwu Lake scenery into bedroom designs in the second floor. The space is open with simple and tidy decorations. The elegant background uses bright dark green and plum yellow to create charms. Living in the guest room can make you gain a cleanness and joy as if being a landscape gentleman and enjoying the beautiful Xuanwu Lake scenery. Living in the master bedroom can warm you body and heart and make you intoxicated as if drinking a pot of plum wine made of mature plums.

"Goose! Goose! Goose! Raising its head and singing to the sky." Kids learning cannot start without the poem in Tang Dynasty *The Geese*. The designer pours more humanistic care into children's room designs, uses tranquil and warm light yellow as the main tone of soft decoration and uses pillows, sailing models and swam stamps full of child interest to foil the atmosphere, which is innocent and beautiful. Outside children's room, the designer makes full use of the original architecture space and skillfully designs a chic family hall, which inherits the essence of metropolitan style. Concise and neat lines and fashionable and comfortable soft decorations present avant-garde sense and exquisite texture of industrial design, which is simple and perfect. The parents can play with their kids and enjoy family happiness.

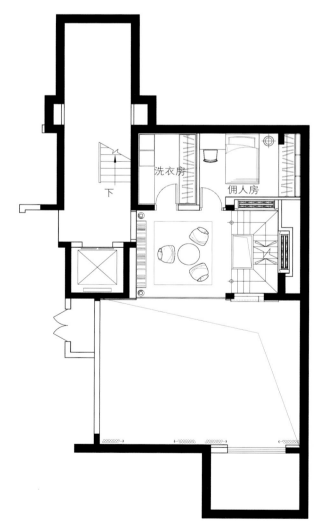

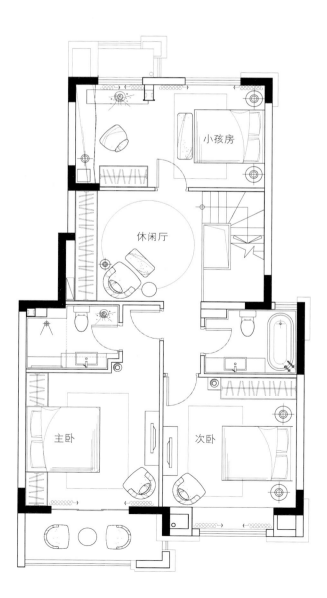

"江南佳丽地，金陵帝王州"。六朝古都南京，拥有2600年建城史和近500年建都史，崇文重教，贵为"天下文枢"，与罗马城并称为"世界古典文明两大中心"，引得无数文人墨客趋之若鹜，形成别树一帜的南朝文化。建筑是死的，但设计是活的，高文安一直提倡有温度的设计，他理性把握南京城历史脉络，解构中西，把文化情怀融入样板房设计，用充满后工业时代前卫感与感性人文色彩的大都会风格，大胆表现歌舞升平的盛世风潮，诠释时尚摩登的现代生活理念。

地下层休闲厅的设计，被赋予浓厚的艺术气息和叫人眼前一亮的空间幽默感，表现派特农神庙遗迹与西欧绅士淑女生活的古典画，由意大利艺术家设计，前卫感十足的云团装饰品，中世纪欧洲风情的青铜兽首，融入中国红色彩元素的时尚家具，透出中西融合的文化底蕴，诠释现代大都会简洁、奢华、雅致的风格与内涵。

"烟笼寒水月笼沙，夜泊秦淮近酒家"。一条秦淮河，流淌着古老南京城千年的风流与雅韵。负一层酒吧的设计，灵感源自艳绝古今的秦淮夜色，跳跃的颜色一如翩翩起舞的金陵十三钗，风情无限；中式陶瓷凳、复活节岛巨人头像金属凳，东西方元素的混搭，描摹出大都会撩人心弦的妩媚性感。

南京古称石头城，石头城最美在清秋，"秋雨梧桐叶落时"，南国女子那种清秀婉约方能尽显淋漓。设计师把石头城秋雨的元素融入楼道设计，高低悬挂的水晶玻璃球，如秋雨从天而降，演绎出秋雨秋风醉居人的动人诗韵。

李白形容南京为"地即帝王宅，山为龙虎盘"，灵秀山川又以紫金山风景最佳。一层生活空间的设计，以紫金山为意向，开放式的客餐厅同气连枝，视线极佳，风景独好。家具的选色上，客厅大红大紫，彰显高贵。餐厅富有视觉冲突之美，最能代表都市人个性又随性的生活主张。

金陵自古出才子，风流才子挥毫泼墨以江山入画。高文安师法古人，以玄武湖风光融入二层卧室设计，空间开阔、装饰精简，素雅背景色里以亮眼的青绿、梅黄造就意韵，居客卧，好似做了那清都山水郎，纵情于玄武湖碧波间，得一份清净欢喜。居主卧，如饱饮一坛熟透黄梅酿的梅子酒，身心俱暖，陶然欲醉。

"鹅,鹅,鹅,曲项向天歌"幼童蒙学离不开唐诗《咏鹅》。设计师在儿童房的设计上倾注了更多的人文关怀,以宁静温暖的鹅黄色系作为软装的主轴色调,再用富有童趣的抱枕、帆船模型、门外天鹅邮票烘托气氛,纯真而美好。儿童房外,设计师充分利用原有建筑空间,巧妙设计了一个别致的家庭厅,继承了大都会风格的精髓,线条简单利落,软装时尚舒适,呈现工业设计的前卫感与精致质感,简单而完美,家长含饴弄孙,尽享天伦之乐。

DREAMLIKE PEACH GARDEN, LEISURE ORIENTAL VILLA

梦回桃花园 悠然东方墅

In this project, the designer breaks, regroups and creates various elements and lines to perfectly combine their luxury and low key, texture and warmth. Edges of the ceiling uses champagne gold metal to outline areas of the living room and dining room, which keeps the connectivity of the space, endows the space with ductility and echoes with decorative metal stripes embedded in the gray wall. Abstract decoration on the background wall presents low-key luxury through design details. The ground is paved with natural gray marbles, which presents rhythm of space and layering of texture. Blue cloth is used in the red brown manual customized retro wooden armchair with old dark leather cushion for leaning on. White leather sofa matches with white marble tea table, which weakens luxurious colors and gives life more jumping colors. In the good rhythm, it is more suitable to live.

Project name: Shenglong • Peach Garden 300m² Villa Show Flat
Design company: Studio HBA
Main materials: marble, cloth, leather soft coverage, etc.

Location: Nanjing, Jiangsu
Area: 300m²

Photographer: Yan Sha

项目名称：升龙·桃花园著300m²别墅样板房
设计公司：Studio HBA
主要材料：大理石、布艺、皮革软包等

项目地点：江苏南京
项目面积：300m²

摄影师：沙岩

In the dining room, the translucent façade of the cabinet uses overlaps, intersections and juxtaposition between lines and surfaces to extend visual planar to three-dimensional space, which creates a subtle "lightness" in the façade.

The master bedroom uses gold to the point and its main characteristic is the breakthrough of material collocation. The bedside background uses large area of cylindrical pattern leather soft coverage. The graphic design with neat lines combines with fashionable rhythm of leather, which deepens modern connotations carried by traditional modeling.

The boy's room injects sculpture forms, art elements and classic toys into vital bright orange. The intervention of unique decorative symbols makes the space present thriving vigor and vitality.

As a recreational space, the underground layer has bedrooms, bar, SPA and gym. The entire tone is deep, low-key and luxurious and presents strong and lively structure lines. Soft tone of the winding chandelier and reflective surface of the organic glass ceiling form a contrast, penetrate into each other and reach the physical transparency, which endows the whole space with various changes. The faint sight percepts metaphor of the space. Neat wood stripes in the bar are embedded in dark marble cabinet, which adds beauty with customized metal furniture with fluent and distorted lines, forming a half-open space. The gym purposely uses gray metal mesh and laser engraving hollow-out, presenting transparent design. The whole wall mirror reflects the whole SPA space and brings sight into the round bathtub with bamboo rattan furnishings. The ground and wall are paved with large area of inclined stripe marbles, which builds construction relation of the space and forms an inner tension. White cylindrical double basin, comfortable massage couch and big gold maple leaf furnishings on the wall perfectly arouse senses and visions.

在这套空间里，设计师将各种元素与线条打碎、重组、新造，将其中的奢华与低调、质感与温馨结合得天衣无缝。天花边缘，用香槟金的金属勾勒出客厅和餐厅区域，又保持空间的连通性，赋予空间延展性，与灰色墙面上镶嵌的装饰性金属条纹交相呼应。背景墙上的抽象装饰从设计细节上表现低调的奢华。地材用天然云灰大理石铺就，表现出空间的韵律与质感层次。蓝色布艺镶嵌在手工定制的具有复古格调的红棕色木质单人沙发，以做旧的深色皮革靠垫加以点缀。白色的皮质沙发，配以白色大理石饰面的茶几，淡化了奢华的色彩，调配给生活更多的跳色，在一张一收的节奏中更加适居。

眼光探索到餐厅，橱柜在半透明立面的处理上，运用线面之间的重叠、交织、并置，将视觉的平面构成延伸到立体的空间中，在立面上创造出一种微妙的"轻"。

主卧走金手法点到为止，更主要的特色在于材质搭配上的突破。床头背景运用大面积柱形图案的皮革软包，线条俊朗的平面设计和皮革材料的时尚节奏互为表里，深化了传统造型框架下承载的现代内涵。

男孩房将雕塑形式、艺术元素和经典玩具融入极具生命力的亮橙色中，因独特装饰符号的介入，使空间诠释出蓬勃的朝气与活力。

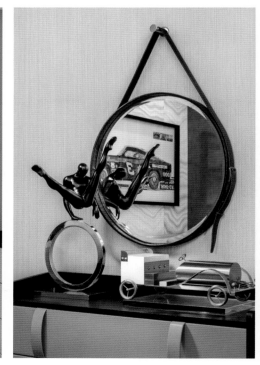

地下一层作为娱乐空间，配备了居室、酒吧、水疗 SPA 及健身房。整体基调以深沉、低奢为主，同时表现强烈明快的结构线条，蜿蜒的枝形吊灯和有机玻璃吊顶，柔和色调与反光表面形成对比，彼此穿透，达到物理上的透明，给整个空间赋予迷离不定的变化，借助视线的隐约，感知空间的隐喻。酒吧规整的木质条纹镶嵌在深色大理石纹柜面中，与定制的具有流畅扭曲线条的金属家具交相辉映，形成一个半开放的空间。健身房特意使用灰色的金属网眼与激光雕刻镂空，表现透光设计。整墙的镜面折射出水疗 SPA 的整个空间，把视线引入以竹藤编织装饰的圆形浴缸中，大面积斜条纹大理石铺就的墙面和地面，建立起空间的构筑关系，形成内在张力。白色圆柱型双台盆，舒适的按摩床，墙面大型金色枫叶的装饰，完美调动了感官和视觉。

HERMES HOME
爱马仕之家

Hermes is the world famous luxury brand, created by Thierry Hermès in Paris, France in 1837 with a history of more than 170 years. It stands in the top of classic clothing brand by its exquisite craft and aristocratic design style. Hermes feature lies on its classics, traditions, multiple and repeated rigorous handicrafts, superior materials and luxurious and noble symbols.

This Hermes themed villa is based on modern style; the space tone is warm gray, brown and white; the desgienr uses concise and vivid method to reinterpert fashion and taste, trying to create an extreme vision and sense of feild. The display color is orange, rose gold, creamy white and brown. The collocation of fresh orange and rose gold is luxurious yet not exaggerated, giving the residents a bold and jumping sensory world. Furnishings, paintings, furniture details and various collocations form a strong visual symbol which is transferred into the space; every element has extraordinary aesthetic taste and forms a kind of temperament which is close to life and obviously beyond life.

Project name: Xixi Zhixin Villa Show Flat
Design team: KNYU TEAM
Location: Wenzhou, Zhejiang
Design company: KNYU CONSTRUCTION
Photographer: Ninglong Xu
Area: 420m²
Main materials: Obama grain marble, blue gold sand marble, orange hard coverage, rose gold stainless steel, tawny glass, etc.

项目名称：西溪置信原墅样板房　　设计师：鲲誉团队　　项目地点：浙江温州
设计公司：鲲誉建设　　摄影师：徐宁龙　　项目面积：420m²
主要材料：奥巴马木纹大理石、蓝金沙大理石、橙色硬包、玫瑰金不锈钢、茶色玻璃等

In the five-floor villa, every space in every floor is placed with Hermes brand elements; every floor has its unique function and corresponding fun. The reasonable space layout makes it a mansion of leisure, entertainment and living. The first floor is relatively open public space; the negative floor has reception room and family video room; the second floor is tranquil bedrooms; the third floor is private space of the host and hostess. Paintings and furnishings wipe out a distinct spark in the new and old collision; calm and wise black, elegant and cool gray tone, bright and enthusiastic orange red and noble and gorgeous rose gold form the temperament colors of the mansion, making the whole space full of gripping beauty.

In the living room, calm and magificent dark color marble and noble and cool warm gray wood veneer are the color foundations, collocating with bright Hermes orange and noble and gorgeous rose gold, which makes the reception place exquisite and luxurious with unique ostentation and extravagance. In the living room and dining room, decorative paintings with Hermes scarves and Hermes carpet match with classic phalaenopsis flowers; orange chairs form a unique scenery; the distinct dining room combines table manners with fashion spirits to create a balanced and elegant dining atmosphere, expressing tastes of the owner and the light luxurious connotations.

The study sets magnificent and calm dark color wood veneer and Hermes hard coverage leather as the main tone; decorative paintings with Hermes scarves and Hermes gray carpet manifest the magnificence. Hermes orange and rose gold luxurious furnishings express the fashionable, individualized and dynamic high-quality lifestyle of modern style, making residents feel the luxurious charms as if in the scene.

Colors of the master bedroom are balanced and full of rich layering with dark color wood veneer as the main tone. Creamy white leather bed background, fresh and bright Hermes scarves decorative painting, calm and low-key black furnsihings, fresh orange gift box and TV backgrond decorative cabinet are fresh yet not dry. Orange, wood color and creamy white in different level are properly uesed in the space, bringing the bedrooms a tranquility and comfort in the powerful and calm tone. The lively enjoyment creates a charming private space for different owners.

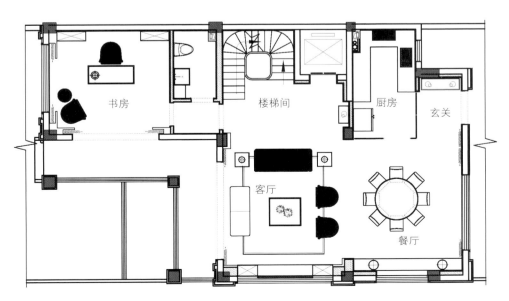

　　爱马仕是世界著名奢侈品牌，1837年由Thierry Hermès创立于法国巴黎，迄今已有170多年的悠久历史。一直以精美的手工和贵族式的设计风格立足于经典服饰品牌的巅峰。爱马仕的特点在于它的经典、传统、手工艺步骤多重反复严谨、材料上乘、有着奢侈、尊贵的象征。

　　本案爱马仕主题别墅是以现代风格作为设计的原点，空间色调以暖灰色与褐色、白色为主调，以简洁明快的手法对时尚、品味重新诠释，力图营造极致的视觉和场所感。陈设色调以橙色、玫瑰金、米白色、褐石色为主，浓艳的橙色和玫瑰金的搭配华贵而不浮夸，给居者一个奔放跳跃的感官世界。同时通过饰品、挂画、家具细节等各处，各种样式的组合搭配，形成强烈的视觉符号转换到空间中，每个元素都蕴藏着卓然的审美品位，形成接近生活又明显高于生活的气质。

　　别墅共五层，每层的每个空间都放有爱马仕品牌元素，空间的每一层都有自己独特功能和对应的趣味。合理的空间布局，使之成为休闲、娱乐、家居为一体的豪宅，一楼相对是比较开放的公共空间，负一楼为会客厅、家庭影音室，二楼是静谧的卧室空间，三楼是男女主人的私享空间。挂画、摆件，新与旧的碰撞，擦出了别样火花；沉稳睿智的黑、优雅冷静的灰色系，璀璨热情的橙红与高贵华丽的玫瑰金，形成这座豪宅的气质色彩，使整个空间扣人心弦的华美。

　　客厅以沉稳大气的深色大理石、高贵冷静的暖灰系木饰面为色彩基调，搭配璀璨的爱马仕橙、高贵华丽的玫瑰金，让这个待客空间精致奢华且独具排场。客厅与餐厅的陈设，爱马仕丝巾装裱的装饰画，爱马仕风格地毯，配以经典蝴蝶兰花卉，橙色的餐椅在空间中形成独特风景，风格独具的餐厅兼融了宴客礼仪与时尚精神，糅合出平衡典雅的用餐氛围，表达了主人品味和轻奢的内涵不谋而合。

书房以绅士大气沉稳的深色木饰面，以爱马仕本色硬包皮革为主调，爱马仕丝巾装裱的装饰画，爱马仕的灰调满铺地毯彰显大气。爱马仕元素的橙色与玫瑰金的奢华摆件，色彩个性，表达了现代风格时尚、个性、活力的高品质生活方式，让居者身临其境的感受爱马仕的奢华魅力。

主卧室空间色彩平衡、层次丰富，以深色木饰面为基调，床背景米白色皮革背景，陈设上以爱马仕鲜艳亮丽的丝巾装饰画，沉稳低调的黑色摆件，鲜艳的橙色礼盒，电视背景装饰柜，艳而不燥。利用不同层次的橙色、木色和米白色，恰当好处地融入空间，在浑厚沉稳的色调中带给卧室里所需的宁静和舒适，空间上的轻快愉悦，为不同的主人营造韵味十足的私密空间。

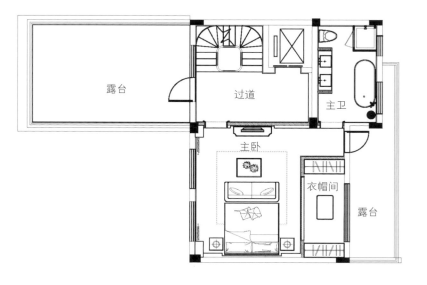

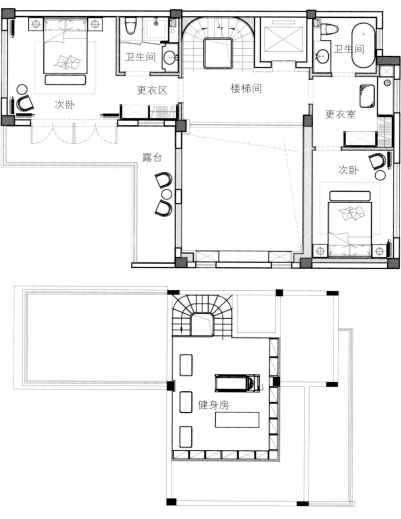

MODERN LUXE
现代奢华

ART FROM LIFE
来源生活的艺术

Located in Beijing, Emerald Park is a representative work of Vanke high-end projects. It has advantageous geographical position, will be created into a livable life city with Boston cultural flavors and is a high-quality area with heavy cultural customs.

Project name: Emerald Park Show Flat
Design company: GBD
Main materials: metal, solid wood parquet, natural stone, etc.

Designer: Bill
Photographer: Bill

Location: Beijing
Area: 250m²

项目名称：翡翠公园样板房
设计公司：广州杜文彪装饰设计有限公司
主要材料：金属、实木复合地板、天然石材等

设计师：杜文彪
摄影师：Bill

项目地点：北京
项目面积：250m²

123

Texture artistic watercolor painting in the living room renders the superior taste of this project and presents the designer's idea to pursue life quality. A set of beige sofa and emerald pillows hold up a comfortable living room and make all colors transition naturally. The elegant dining room presents skillful color collocation and intersection of lines. Detail arrangement creates a unique life situation which is exquisite and elegant. Refined plants bring the whole atmosphere a sense of aura. Under the basic tone with white color and wood texture, the flower art room is decorated with charming blue, warm flowers and plants; the proper color as if a finishing point makes life interesting. Metal materials and exquisite texture in bedrooms are skillfully used, making the space full of layering. With a small lamp open, every corner has its own stage; the crystal clear soft lights create a three-dimensional sense. Under careful arrangements of the designer, every piece of furniture emits its own charm in its corner, which is harmonious and creates a living space full of strong artistic atmosphere. Customized decorations present exquisite originalities, give people an elegant texture and delicate style and gain perfect artistic embodiment.

　　翡翠公园位于北京，是万科高端项目代表作，其地理位置优越，将打造为极富波士顿人文气息的宜居生活城，是一个文化风情浓郁的优质区。

　　客厅匹配带有质感的艺术水彩画，渲染了本案的卓越品味，体现了设计师对追求生活质量的想法。一组米色系的沙发及祖母绿色彩的抱枕撑起了舒适的客厅空间，让所有可被发挥的色彩都能自然过渡。高雅气息的餐厅，体现在色彩的巧妙搭配与线条绝妙的交错感。细节的布置营造出独特的生活情境精致又不失典雅，清雅的植物点缀，给整个氛围带来一丝灵气。花艺室在白色与木质铺陈的空间基调下，妆点着些许沁心的蓝、暖心的花与绿植，恰到好处的用色，如同画龙点睛的神来之笔，让生活更有味。卧室金属的材质与细腻的质感被巧妙地运用，让空间凸显更多层次。开着小灯，每处角落都有属于自己的舞台，衬以晶透柔和的光效，制造了空间的立体感。每一件家具在设计师的精心安排下，在各自所在的角落里散发着独特的魅力，又互相和谐，创造了一个艺术氛围极强的家居空间。量身定造的装饰尽显考究匠心，给人一种优雅质感和精致风尚，获得了完美的艺术体现。

POETRY · METAMORPHOSIS
诗意·蜕变

Soft decorations of this project use many methods to interpret themes of "nature", "ecology" and "livability" and inject outdoor raw stone, collectable fossils and crystal stones into the whole design, which makes the project perfectly combine with luxury and nature and is called as the most poetic villa show flat. The application of butterfly elements becomes another featured theme of this show flat. Butterfly symbolizes happiness and love, which gives people inspiration, intoxication and yearning.

Project name: Great Blue Mount Villa
Design company: GND DESIGN LIMITED N+
Designer: Rui Ning
Location: Dongguan, Guangdong
Area: 160m²

项目名称：卓越蔚蓝山别墅
设计公司：GND设计集团——N+恩嘉陈设
设计师：宁睿
项目地点：广东东莞
项目面积：160m²

The ancient Indian legend tells us that the couple or the blessed one tells their wishes to butterfly in hand and lets it go; then the butterfly will tell your wishes to the angel; the angel will witness your love and everlasting promise and make them last forever and fulfill your wishes. Butterfly is also Chinese traditional mascot; it only has one lifetime soul mate and is the representative of faithful love. Butterflies who love flowers are used to symbolize sweet live and happy marriage. Butterfly is also the representative of peace, freedom, happy life and auspiciousness and reflects people's pursuit of perfection.

Lamps, paintings and furnishings use butterfly elements many times, which better interprets messages conveyed by this project in effect and implication. The daughter who loves collecting herbariums hangs her favorite works in the bedside. Room of the young son who admires various heroes has ubiquitous heroism. Designs and presentations of these portraits better foil the life state of the family. The elder's room is tranquil and comfortable; the master bedroom is graceful and luxurious; the collectable paintings manifest the taste and temperament of the owner.

本案软装设计上运用多种手法去诠释"自然"、"生态"和"宜居"的主题,将户外的原石、收藏级的化石、水晶石等元素引入到整个设计,使得项目很好地结合了奢华与自然,被称为史上最有诗意的别墅样板房。蝴蝶元素的应用,成为该样板间的另一特色主题,蝴蝶本身寓意着幸福、是爱情的象征,它给人以鼓舞、陶醉和向往。

古老的印地安传说告诉我们,新人或者祝福的人把自己的心愿轻声告诉手中的蝴蝶,然后将蝴蝶放飞,蝴蝶就会把你的心愿告诉天使,天使将见证你爱的讯息和天长地久的承诺,让它恒久不变,实现你的愿望。蝴蝶也是中国传统的吉祥物,蝴蝶一生只有一个伴侣,是爱情忠贞的代表,恋花的蝴蝶常被用于寓意甜蜜的爱情和美满的婚姻。蝴蝶还是和平、自由、幸福生活、吉祥美好的象征,表现了人类对至善至美的追求。

灯具、挂画、饰品中多次应用蝴蝶的元素,在效果和寓意上都很好地诠释了此项目要传达的讯息。喜爱收藏植物标本的女儿,将自己最喜爱的

作品悬挂床头;崇拜各种英雄的小儿子房间中,随处可见的英雄主义;这些人物肖像的设计及展现都更好地衬托了家庭的生活状态。老人房的宁静与舒适;主人房的雍容华贵,收藏级化石的挂画,都凸显主人的品位与气质。

MODERN LUXE 现代奢华

LIGHT LUXURIOUS ELEGANCE, LIVING AESTHETICS
轻奢雅致 人居美学

If you cannot go to Canada to appreciate the beauty of ice lake, just come to Huangjiang Town to feel the beauty of natural residence. Located in the core city center of Huangjiang Town, this project has convenient transportation and attracts many customers from Shenzhen and other surrounding cities. Hard decoration of the project is positioned as "Hong Kong style" and its soft decoration is designed by our company. Collocating with luxurious interface of hard decoration, the soft decoration uses crystal, fur, leather and bright texture furnishings to strengthen the luxurious and elegant theme. Fashionable and comfortable soft decoration presents noble feature of the owner. The main tone is metal color, collocating with classic black, white and gray and injecting with coffee, beige and dark red, which is luxurious and fashionable. Exquisite and elegant furnishings present high-end position of the project and status symbol of the owner while partial details present personal taste and interest of the owner; every corner emits a supreme quality.

Project name: Dongguan Banff Spring Villa Show Flat 20#
Design company: Steven Lee & Partners Interior Design Co, Ltd
Designer: Chao Lee
Location: Dongguan Guangdong
Area: 309m²

项目名称：东莞班芙春天20#别墅样板房
设计公司：深圳市圣易文设计事务所有限公司
设计师：李超
项目地点：广东东莞
项目面积：309m²

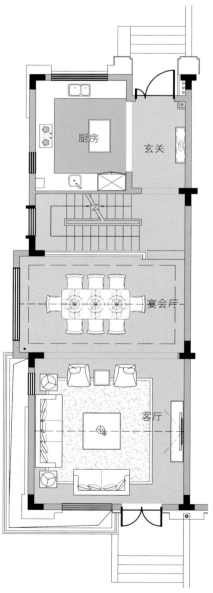

如果不能去加拿大领略冰原湖光的美，来黄江就可以感受自然人居的美好。本案处于黄江新城市中心的核心位置，交通便利，吸引了来自深圳等周边城市顾客的青睐。项目的硬装定位为"港式风格"，由我司配合软装陈设设计。结合硬装奢华的界面，软装通过水晶、皮毛、皮革、亮面质感的家饰强化奢华、优雅的主题。通过软装时尚舒适的搭配，凸显客户的尊贵身份特点。金属色为主，搭配经典的黑白灰，注入咖色、米色、暗红色，奢华而不失时尚。精致优雅的装饰，体现项目高端的定位和主人的身份象征，而局部的细节体现出主人的个人品味和爱好，每一个角落散发出至高的品质。

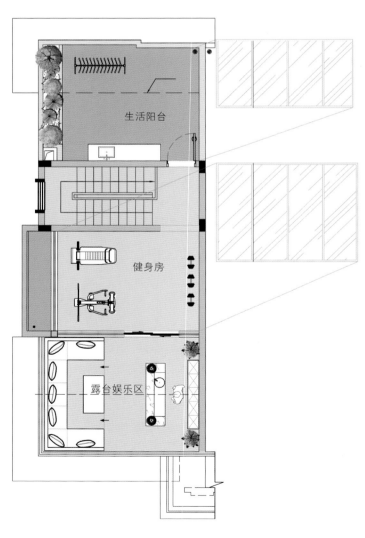

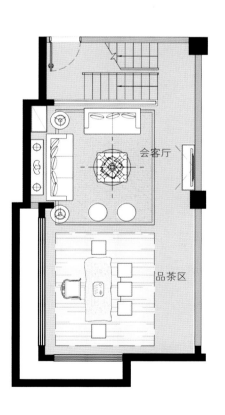
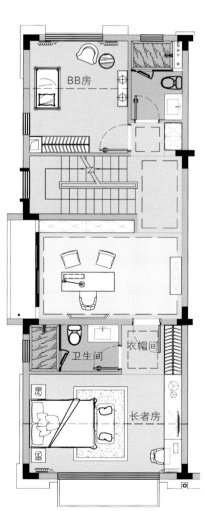
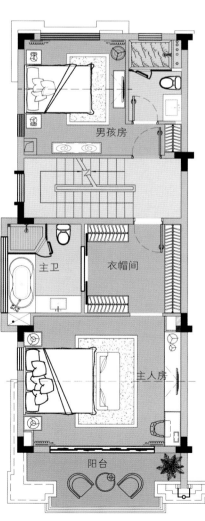

FASHION AND CONCISENESS IN ROUND AND SQUARE

于方圆之中 见时尚气质

The space structure uses "square and round" throughout the space, which manifests modern unique craft modeling and stresses on creating details. The combination of straight lines and curves uses the texture of the metal, the thickness of stoving varnish and the gloss of silk to create unique style taste of the owner. This is free, open and desirable. The use of classic sofa and its redesign and re-creation endow the entire space with new life, which is also the self-interpretation and yearning of high quality life. The dining room and living room are connected together, which is transparent, bright and smooth on the whole decoration. The master bedroom uses gold and gray as main tones; exquisite color collocation brings the space texture enjoyment. The subaltern room mainly uses gray; the bedside background is decorated with gold furnishings, which makes every space echo with each other. The children's room chooses avant-garde wallpaper, which conforms to vigorous vitality of the children.

Project name: Shenzhen Unicenter Jiuyu Show Flat House Type C
Design company: MATRIX
Main materials: gold lacquer board, marble, wallpaper, wood veneer, gold stainless steel mosaic, wood floor, etc.
Chief designers: Pengjie Yu, Guan Wang, Jianhui Liu
Location: Shenzhen, Guangdong
Area: 85m²

项目名称：深圳壹方中心玖誉C户型样板房
设计公司：矩阵纵横
主要材料：金色烤漆板、大理石、墙纸、木饰面、金色不锈钢马赛克、木地板等
主创设计师：于鹏杰、王冠、刘建辉
项目地点：广东深圳
项目面积：85m²

　　空间构造以"方与圆"相互穿插运用，凸显现代独特的工艺造型，讲究细节的打造，在直线与曲线的交合之中，用金属的质感、烤漆的厚度、丝质的光泽打造出主人特别的风格品味。这是自由的、开放的、也是令人向往的。沙发经典款型的运用，与再设计再创作，赋予整体空间新的生命，也是对高品质生活的自我诠释和向往。餐厅和客厅相连，通透明亮，在装饰上一气呵成。主卧以金色和灰色为主色调，精致的色彩搭配给空间带来更富质感的享受，次卧以灰色为主，床头背景墙以金色挂件为装饰，让各空间之间相互呼应，儿童房则选择前卫的壁纸装饰，契合孩子年轻的蓬勃生机。

IDEAL ART WITHOUT BOUNDARY
理想的无界艺术

Themed as ideal art without boundary, this project reveals the designer's pursuit of perfect art with modern decoration design method; the collision between delicate and exquisite decoration and art works presents an elegant retro art different from luxurious art.

Tall height of the living room and modern circular droplight make the space taller, collocating with artistic furnishing articles, which creates a free and individualized atmosphere. Droplight spliced by metal tubes and elegant and fluent lines in the dining room make the space full of layering; the collision of weaving art and metal material and art sparkles produced by different elements endow the active style with emotions. The master bedroom continues modern and elegant style in the living room; calm neutral tone matches with droplight with flamboyant lines, which is concise yet not simple; warm and harmonious colors manifest noble tone and set off each other. The swift horse of the talent scout and artistic derivatives become new favorites of bedroom design. Orange in subaltern room is the symbol of fashion and luxury and the representative of passion and vitality; your exclusive instrument is defined by yourself. Well-organized decorative shelf and geometric splicing carpet are individualized and decent.

Project name: Park Show Flat One
Design company: GBD
Main materials: wood floor, natural stone, metal, etc.

Designer: Bill Du
Photographer: Bill

Location: Beijing
Area: 230m²

项目名称：公园里样板房壹
设计公司：广州杜文彪装饰设计有限公司
主要材料：木地板、天然石材、金属等

设计师：杜文彪
摄影师：Bill

项目地点：北京
项目面积：230m²

　　本案以理想的无界艺术为主题，透露出设计者对完美艺术的追求，配以现代装饰设计手法，精致细腻的装饰与艺术品之间的碰撞，呈现出有别于奢华的典雅复古艺术。

　　客厅挑空的高度，现代风格的环形吊灯，让整个空间更加高挑，搭配富有艺术性的装饰摆件，营造出自由、个性的氛围。餐厅以金属管拼接的吊灯，高雅流畅的线条，使空间层次更具层次，编织艺术与金属材质相互碰撞，不同元素之间产生的艺术火花，让活跃的格调为之增加情感。主卧延续了客厅现代典雅的风格，沉稳的中性色调配以线条张扬的吊灯，简约而不简单，温暖和谐的色调中透着高贵格调，相互映衬。伯乐的千里马与艺术的衍生品，成为卧室设计的新宠。次卧橙色是具备时尚奢华的符号，更是激情与活力的代表，你的专属乐器由你定义。错落有致的装饰架子与几何拼接的地毯，个性又不失大方。

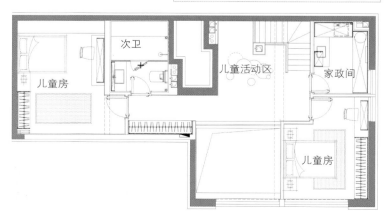

MODERN LUXE | 现代奢华

WHEN CALMNESS AND GENTLENESS MEET BY CHANCE

当沉稳与温婉不期而遇

When life is immersed in style, though it is quiet, its temperament is showed through behavior and expression. Even the passing moment can leave a deep impression. When restraint and luxury meet by chance, tastes are revealed naturally. Collecting various respectful spaces only needs to cultivate a kind of unique taste, which is implicit, inspiring, luxurious yet not flamboyant.

Project name: Huaihua Royal View Villa
Design company: BedA Garden
Main materials: solid wood, metal, leather soft coverage, etc.

Designer: Rong Si
Photographer: YUANSHENG STUDIO

Location: Huaihua, Hunan
Area: 400m²

项目名称：怀化帝景别墅
设计公司：百搭园软装
主要材料：实木、金属、皮革软包等

设计师：司蓉
摄影师：建筑空间摄影

项目地点：湖南怀化
项目面积：400m²

Solid wood materials in the living room are natural and noble. The designer uses simple European design technique with Chinese connotations to create a "gentle and beautiful" space. Through perceptual extraction, rational processing and application, the designer remains time and uniqueness of the space. The style of the living room is calm with retro sense from copper. The entire collocation has firm edges and exquisite temperature, which is soft and strong. Concise and neat lines of chaise longue and tea table match with exquisite handmade leather and restrained and warm carpet. Leaning on the sofa, the orange presented in front of your eyes can make the mood jaunty and clear.

Tasting the flavor of life, we can start from the kitchen and just enjoying cooking and food. In good mood, pots and pans are as ingenious as music instruments and can also play wonderful music.

In the elder's room, restrained and heavy furniture collocates with warm wood materials, making the whole space natural and simple with high quality. Exquisite and elegant pendant furnishings at the bedside are warn and aesthetic, presenting quality life, where the elder can enjoy life independently.

The master bedroom derives tastes from details. Leisure, comfortable and individualized decoration style and perfectly interpreted tone and furnishings endow the space with free and magnificent atmosphere.

Self-restrained space contains the calmness of gentleman and gentleness of mature lady who only pursue comfort. In calm and tranquil atmosphere, command and manner come together; the gestures and expressions can make a difference. When succeeded, views and support from others are less important than inner leisure. Let go of the distractions and pick a pot of green tea, the mind can be free in the fragrance tea environment.

当一种生活沉浸在格调之中，虽不动声色，气质已然流于行表。即使在擦肩而过的一瞬，也能让人印象深刻。当内敛与华贵不期而遇，品味便自然流露出来。收纳各式尊崇空间，只需培养一类独特的品味，既含而不露，又隐而不发，富贵而不张扬。

客厅的实木材质自然而高贵，设计师运用简欧和具有中式底蕴的设计手法为其营造出一处"温婉至美"的空间。经过感性提取、理性加工与运用，保持了空间的时代性和独特性。客厅风格沉稳，加入了金属铜的复古感。整体搭配上，有坚定的棱角，也有细腻的温度，柔和却有力。贵妃椅与茶几简明干脆的线条，交映着手工皮革的精致与搭毯内敛的温和。倚靠在沙发上，呈现在眼前的那一抹橘黄，有时能让心情不由得雀跃、明朗起来。

品味生活的味道，我们可以从厨房开始，乐享烹饪及美食。心情极好之时，锅碗瓢盆仿佛乐器般巧妙，也能奏响美妙的乐章。

长辈房内敛而厚重的家私款式搭配温润的木质，让整个空间自然而淳朴又不失品质。床头精致高雅的挂饰，独立享受，温馨唯美，体现品质生活。

主卧从细节衍生出品味，悠闲、舒畅、个性的装饰风格，完美诠释的色调与摆设，让空间呈现自由、瑰丽的氛围。

涵养之所，蕴含绅士的沉稳抑或娴熟女性的温婉，谈笑间惟求舒适。风平浪静之下，指挥与气度相博，举手投足间，力挽狂澜；当功成名就信手拈来之时，旁人的仰视与簇拥已非重要，而今，更在意内心之从容。放下纷扰，佐一壶清茶，在茶香的缭绕沁润中，心智无所不达。

MODERN S

MPLICITY

现代简约

明快柔美的线条，富有节奏感，注重整个立体形式与室内外的沟通，竭力给室内装饰艺术引入个性新意，整体风格大气，简约而不简单。

Lively and mellow lines with rich rhythm, highlighting the entire three-dimensional form and indoor and outdoor communication, trying to introduce new personality into interior decoration art; the entire style is magnificent, concise yet not simple.

THE BEAUTY OF BALANCE
平衡之美

MODERN SIMPLICITY ■ 现代简约

The interior design of the whole space is inspired by the noble tribute to Art Deco Movement. The designer combines some California style featured by the architectural structure itself to present artistic aesthetics and modern sense of the space through extraordinary Art Deco. France is the birthplace of Art Deco. In the design process, the designer reviews some most representative architectural and artistic elements in its 20s and 30s when Art Deco was the most prevalent and French elegance was popular, absorbs inspiration from them and combines his ingenious conceptions to inject art into interior design.

Project name: Shenzhen Penthouse
Design company: Dariel Studio
Designer: Thomas Dariel
Location: Shenzhen
Area: 860 m²

项目名称：深圳顶层公寓
设计公司：Dariel Studio
设计师：Thomas Dariel
项目地点：深圳
项目面积：860 m²

For the designer, it not only is an ideal residence full of vitality and life, but also conveys his artistic and elegant lifestyle. Every space in the penthouse duplex is filled with a living environment with love to art. Bright and strong colors, decorative surface ornamentations and asymmetric lines and shapes bring a quaint and funny atmosphere.

To better interpret art in the space, these artistic elements not only reflect on walls, floors and ceilings. Our customized furniture and furnishings make our thoughts about Art Deco being expressed better. We believe that to design a space is not only to make the space full of imagination, but also to inspire people around it. This elegant and tranquil residence with culture and art can attract those who have unique tastes and visions of life. They may appreciate paintings by Koloman Moser, Kay Sage, Francis Picabia and Ettore Sottsass as us.

At the same time designs focus on the coordination of surrounding environments. Located in the pentohouse of a tall building, the elegant deplex residence frees from the noise and complication of the city and gains calmness as if getting rid of the earthly world and low-key and luxurious life colors in turn. The designer wants to convey this kind of life concept through design. Repeatedly geometric lines in the space, elegant and profound blue and green are like the scenery outside the window; green mountains and blue sky enhance each other's beauty. These details in the space express a calmness and an elegance.

The space created by the designer as a Parisian not only presents Art Deco esthetics but also infuses Paris elegant tastes in the whole design process. The stylish contemporary works of art occasionally intersperse the entire space, bringing an unparalleled artistic breath and reflecting the life attitude of loving art and design. Original design technique, distinct visual effect, demanding control of details, furniture and lamps from European top designers and high-quality humanistic equipments create a unique yet not flamboyant beauty of balance.

　　整个空间的室内设计灵感源于对装饰艺术运动崇高致敬，结合建筑结构自身特有的一些加州风格，通过超凡的装饰艺术展现空间的艺术美学以及现代感。作为装饰艺术的发源地，法国，在设计过程中，设计师回顾了其在二三十年代，这些艺术史上装饰艺术最盛行以及法式高雅风靡的年代中最具代表的一些建筑及艺术元素，从中汲取灵感，结合自身的巧妙构思将艺术融入进室内设计中。

　　对于设计师而言，这里不仅仅是一处充满活力，生意盎然的理想居所，他更是传递着艺术及高雅生活方式，在顶层复式的每个空间，都弥漫着一种对艺术充满热爱的生活环境。明亮强烈的色彩，装饰性的表面纹饰，不对称的线型和形状都蓄意带来一种奇特而有趣的氛围。

　　为了让艺术在空间中得到更好的诠释，这些艺术元素不仅仅反映在墙面、地面及天花上，我们亲自设计定制的家具以及如地毯等饰品，使得我们对于装饰艺术的想法能得到更好的表述。我们也相信，设计一个空间，不仅仅是让空间本身充满想象，同时它应该能让周遭的人获得更多灵感才是。如此一间集文化与艺术于一身的雅致幽所，定能吸引对于生活有着独到品味和眼光的人们，他们可能也会与我们一样欣赏着 Koloman Moser 的画作，亦或是像 Kay Sage，Francis Picabia 或者 Ettore Sottsass 这些艺术巨匠。

　　设计的同时，非常注重于周围环境的相互协调。坐落于高楼顶层的复式雅居，脱离着城市应有的喧嚣繁杂，反之获得的是如同摆脱尘世的安然自得，低调却又不失华丽的生活色彩。设计师希望能够将这样一种生活理念通过设计传递。空间中反复出现的几何线条，高雅深邃的蓝色及绿色宛如窗外景色一般，青山、蓝天，互相辉映。这些存在于空间中的点滴细节，无一不诉说着一种沉静，一种雅致。

　　同样，作为一个巴黎人，设计师创作的空间中不仅仅完全呈现了装饰艺术美学，在整个设计过程中他也融入了巴黎式的高雅品味。而风格化的当代艺术作品不时地点缀着整个空间，带来了无与伦比的艺术气息，也更好地反映了热爱艺术和设计的生活态度。个性独到的设计手法，鲜明的视觉效果，对于细节的苛求把控，融入欧洲顶尖家私及灯具设计师作品以及高品质人性化设备，营造出独特而不张扬的平衡之美。

MODERN SIMPLICITY
现代简约

A CONCISE HOME
简约之家

"I would like to plunge deeply into life, suck marrows of it, live a solid and simple life, get rid of everything beyond life and push life to the matchless place by the most basic form, simplicity, simplicity and simplicity."

——*Walden* by Henry David Thoreau

Interior design: K2LD
Display design: Eric Tai Interior Design
Soft decoration: Eric Tai Interior Design
Main materials: Glex gray marble, snow dolomite, walnut, white latex paint, etc.

Designer: Eric Tai
Photographer: Weizhong Chen

Location: Shenzhen, Guangdong
Area: 498m²

室内设计：新加坡K2LD建筑设计事务所
陈设设计：深圳戴勇室内设计师事务所
软装定制：深圳戴勇室内设计师事务所
主要材料：格力士灰云石、雪花白云石、胡桃木、白色乳胶漆等

设计师：戴勇
摄影师：陈维忠

项目地点：广东深圳
项目面积：498m²

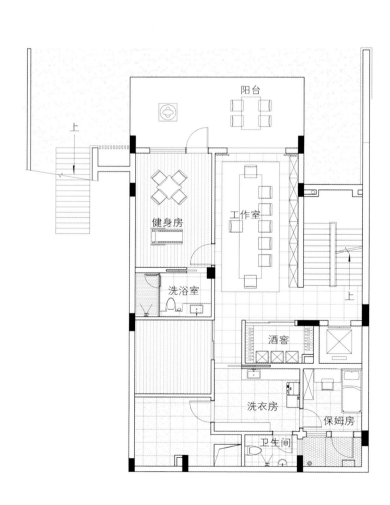

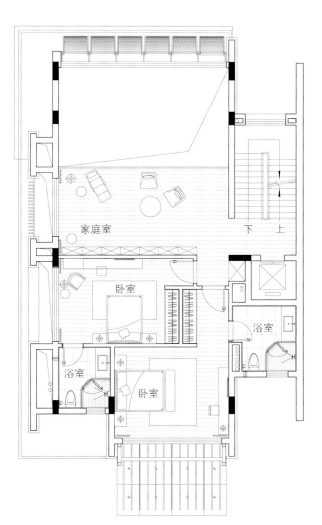
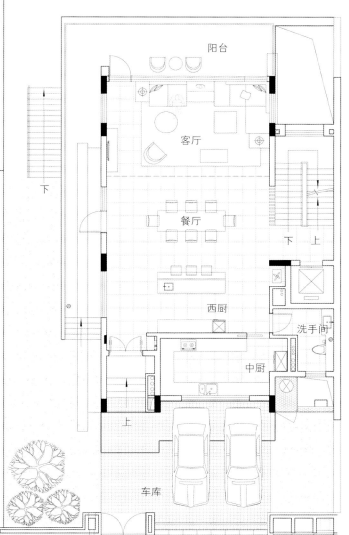

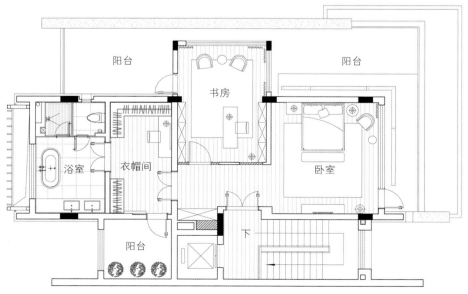

Located in the core area of Shenzhen Bay business circle, this project is a tranquil and elegant residence in the busy city. Purest space and concise furniture deposit in the soft and changeable lights. Being here, tranquil and dustless space temperament is around you, where you can be absorbed in the rhythm of the space to realize spiritual indifference and leisure.

This concise home breaks formal layout and has virtual and real progressive designs through well-organized space planning. The designer Eric Tai suggests the owner to change the original beige stone into Glex gray marble on the first floor and the negative floor when conducting the furnishing designs. Dark gray ground and clean white wall form a distinct contrast. He also changes the original standard western kitchen island into snow dolomite, which strengthens texture changes of materials. Beige hemp spherical lamp furnishings are hung above the living room; exquisite and dark elegant beige leather sofa and dark coffee tea table have different textures and echo with each other. From space to dark and light collocation of furniture, lights shed into interior through tall French windows, which foils natural walnut with coffee grains and presents Zen-like Oriental humanistic flavor and plain freedom.

Stepping up, a whole wall white oak bookshelf in the second floor staircase creates a peaceful reading space; modern furniture combines cloth with leather, making the whole space layout fluent and natural. Zen landscape ink painting created by artist Zhou Zhouzhou presents sparse and lofty, elegant and vacant, clear and clean artistic conception, which is concise and free. Concise Spanish Nomon wall clock, random and exquisite sand stone sculpture and banyan and cypress, smooth wood grain and stone grain have skillful inner connections, creating a space image with rich and harmonious layering. Savoring carefully, you can understand every perfection and measurement in this concise home and feel every subtle ingenuity.

The bamboo shadows sway with the wind; the water is quite and tranquil; the fresh orchid fragrance makes you forget the mortal world. In such a concise and elegant house, every step forward, there is a different scene. After the vanity and noise fade as dust and bloom lightly, all is natural. There are so many sceneries in lifetime while non is comparable to inner freedom and relaxation.

"我愿意深深地扎入生活，吮尽生活的骨髓，过得扎实，简单，把一切不属于生活的内容剔除得干净利落，把生活逼到绝处，用最基本的形式，简单，简单，再简单。"

——梭罗《瓦尔登湖》

本案位于深圳湾商圈的核心位置，是隐于繁华都市中的静居雅境，纯粹到极致的空间与极简的家具沉淀在柔和变幻的光线中，置身其中，静谧无尘的空间气质环绕周身，沉醉在这空间的韵律中体悟心境的恬淡与闲适。

信步而上，二层楼梯间一整面白橡木书架辟出一隅静心阅读之所，依旧是布艺与皮质结合的现代极简风格的家具，整个空间布局流畅自然。由艺术家周洲舟精心创作的禅意山水墨品绘出的疏淡高远、清逸空灵、净澈澄明的意境，简练洒脱，西班牙 Nomon 极致简约的挂钟，随意精致的沙盘石雕与榕柏，细密流畅的木纹与石理，各饰物之间结成巧妙的内在联系，构成层次丰富和谐的空间意象。细细品味，方能读懂简约之家的每一处的极致与揣摩，感受到每一丝匠心独运的微妙。

竹影扶风，白水静穆，一阵清新的兰香足以让人忘却尘俗。身居简约雅舍，随走随望，移步换景，止于心安。在浮华与喧嚣如尘埃般散尽后，淡淡地绽放，一切自然而然。人生多少风景，终不抵内心的自在和轻松。

MODERN SIMPLICITY
现代简约

STROLL IN FOREST
丛林漫步

In the noisy city, we face with all kinds of pressure and annoyance every day, we walk faster and farther and we need a pure land and a place to free hearts to stroll in forest, stay with nature and see the deer under a group of butterflies.

Project name: Zhongzhou Tianyu Second Stage Show Flat House Type E-1
Design company: HORIZON SPACE
Chief designers: Song Han, Qisheng Yao, Shiru Li
Main materials: light color water wood veneer, white stoving varnish wood veneer, Eurasian wood grain marble, advanced artificial stone, jazz white marble, agate jade, wired copper stainless steel, gray lens, etc.
Location: Huizhou, Guangdong
Area: 144m²

项目名称：中洲天御二期样板房E-1户型
设计团队：昊泽空间设计有限公司
主创设计师：韩松、姚启盛、黎师茹
主要材料：浅色水磨木饰面、白色烤漆木饰面、欧亚木纹大理石、高级人造石、爵士白大理石、玛瑙玉大理石、拉丝铜不锈钢、灰镜等
项目地点：广东惠州
项目面积：144m²

Based on this kind of spiritual pursuit, we make the space layout as transparent as possible so as to make people walk freely in it. The materials use light color wood veneer and white stone to create a warm, natural and comfortable space feeling. Entering the space, what presented in front of you is "a deer under a group of butterflies" with stage effect. The white deer is lose in the forest when strolling and is absorbed by the dancing butterflies; it stops to look up quietly and forgets about time. Moving forward is the dining room with mild fun. The ceiling deliberately creates a feeling that a skylight sparkles through the grilling, making people feel outside when dining. The neighbor is capacious and comfortable living room. The painting of its sofa background wall is from a scene in the documentary Dust and Snow, giving people a peaceful and serene feeling. The whole artificial stone wall which connects the living room and dining room is simple and full of sense of detail and layering. Its extremely long size brings people strong visual shocks. 75-inches television and wired copper stainless steel layer and low cabinet form a funny picture, in addition with the sculpture The Walking Man from Giacometti to express the yearning of getting away from the city and returning to nature. In the balcony surrounded by green, with a big lounge, a small side table and a wisp of fragrant coffee, life can be slowed down, leisure and comfortable.

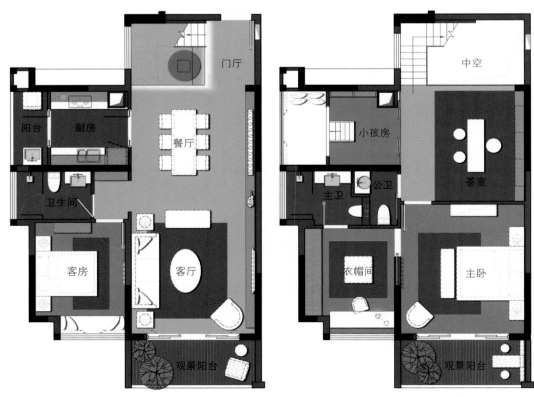

在喧嚣的都市中，我们每天都面对着各种各样的压力和烦恼，我们走得越来越快，越来越远，我们需要一方净土，一个放逐心灵的地方——漫步丛林，与自然为伍，麋失蝶群。

　　基于这样的精神诉求，我们在空间格局上尽可能的通透，以期人能在其中自在游走。在材料上，使用浅色木饰面和白色石材营造一种温暖、自然、舒适的空间感受，进入空间后首先映入眼帘的便是极具舞台效果的"麋失蝶群"——白鹿在林中漫步的时候被翩翩起舞的蝴蝶吸引了，它停下了脚步，抬头静静的望着，竟这样忘记了时间。再往前是极具野趣的餐厅，天花造型刻意营造出一种天光透过格栅洒进来的感觉，让人在用餐时感觉置身室外，与之毗邻的是宽敞舒适的会客厅，主沙发背景的画作是源自纪录片《尘与雪》中的一帧画面，给人宁静安详的感觉。连通客餐厅的整面人造石墙面，简单而充满细节和层次感，超长的尺度给人带来强烈的视觉震撼，75英寸的大电视和拉丝铜不锈钢的层板和矮柜形成有趣的画面关系，再配以贾科梅蒂的雕塑——《行走的人》，来表达我们想要逃离都市，回到自然的渴望。在被绿色包围的阳台上，一把大躺椅、一个小边几、一缕咖啡香，闲适的、慵懒的，生活就这样慢了下来。

MODERN SIMPLICITY ■ 现代简约

TRANQUIL AND WARM SUNSHINE

静谧暖阳

The space sets warm wood color and elegant beige as the main tone; orange tables and chairs bring active and vivid atmosphere; elegant and tranquil blue beddings are comfortable and pleasant. The entire space atmosphere is tranquil and peace as if bathed in the warm winter, which is leisure and free. As the designer says that some things and objects are always here while you cannot enjoy them, for example, the tranquil and warm sunshine. "Some people don't know each other well, though they meet every day; some people know each other well, thought they have less relations." What the designer's work brings is not only aesthetics but also understandings of life.

Project name: Longfor Spring River Shore
Design company: FLY Decoration
Designer: Xiongfei Song
Photographer: Song Ye
Location: Hangzhou, Zhejiang
Area: 198m²
Main materials: cement brick, coating plate, solid wood floor, art paint, stoving varnish plate, KD board, wallpaper, etc.

项目名称：龙湖春江彼岸
设计公司：菲拉装饰
设计师：宋雄飞
摄影师：叶松
项目地点：浙江杭州
项目面积：198m²
主要材料：水泥砖、涂装板、实木地板、艺术涂料、烤漆板、KD板、墙纸等

空间以温暖的木色和素雅的米色作为空间的主色调，点缀的橙色桌椅带来活泼灵动，幽雅静谧的蓝色床品舒适宜人。整个空间氛围静谧平和，如同沐浴在冬日的暖阳中慵懒自在。如同设计师所说，有些东西，事与物，它一直都在，你却不能好好地享有，就像这安静温暖的阳光。"有些人，即使是天天相见也不会相知；有些人，即便是相交甚少也彼此了解"。设计师作品带来的不光是审美，还有对生活的体悟。

FOREST HOME
森林之家

MODERN SIMPLICITY — 现代简约

The lake, skiing children, breakfast under the birch tree, green grass...
This is the home of Carl Larsson.
To be a person like Carl,
Bring your most beloved wife, children and family, to go to the Yukino Lake...
There are lotus all around Daming Lake with poplars and willows in three sides; half of the mountain scenery is reflected in the water.

Life here is like Carl Larsson's painting.
Home, a place to start.
Now, please close your eyes quietly...
Their house is near the lake and surrounded by forest.
As if in the end of the fairy tale world, we realize what is the breathing house. Concise designs, jumping colors, comfortable sunshine and so on, all the atmosphere is just right.

Project name: Jinan Yukino Lake Flat
Design company: Xisheng Architectural Design Studio
Main materials: wood, cloth, wallpaper, etc.
Designers: WenZhe He, Yaohua Tang, Chenwei Gu
Photographer: Shenfeng Zhu
Location: Jinan, Shandong
Area: 85m²

项目名称：济南雪野湖平层
设计公司：西盛建筑设计（上海）事务所
主要材料：木质、布艺、壁纸等
设计师：何文哲、唐瑶华、顾陈玮
摄影师：朱沈锋
项目地点：山东济南
项目面积：85m²

Large French glass window links up the interior and exterior with plentiful sunshine and beautiful natural landscape. Whether forest or lake scenery, all can give residents best enjoyment. The living room uses wood as the main material, which integrates better with nature; the big tall ceiling makes interior broader. Shallow yellow and light blue combine comfort with fashion and endow the space with more charms. The piece of splicing Sudoku painting on the background wall with light gray tone and outstretched grains brings guests strong visual tension. We love nature and enjoy fashion. All people-oriented design should be loved by people.

那片湖，滑雪的孩子、白桦树下的早餐、青青的草地……
这里是卡尔·拉森的家。
做一个像卡尔一样的人，
带着你最心爱的妻子、孩子和家人，去往雪野湖……
四面荷花三面柳，一城山色半城湖。
这里的生活场景就像卡尔·拉森的画。
家，一个开始的地方。
现在，请你轻轻地闭上眼睛……
他们的房子就在湖边，四周是森林。
仿佛身处童话世界的尽头，我们体会到了什么是会呼吸的房子。极简的设计，跳动的色彩，舒适的阳光等等，一切的氛围是如此刚刚好。

超大的落地玻璃墙，将室内与室外零间隔互动，不仅阳光充沛，自然景观也是极佳。无论是森林还是湖景，都能给居住者最完美的享受。客厅以木质为主要材料打造，很好地与自然融合，大大地挑高层次，让室内更加开阔。而浅浅的黄调与淡淡的蓝调，将这种舒适与时尚巧妙结合，赋予空间更多的魅力值。最爱背景墙上那幅拼接的九宫格挂画，浅灰的色调，伸展的纹路，带给宾客极强的视觉张力。我们热爱自然，同时也享受时尚。一切以人为本的设计，都该受到人的喜爱！

DREAM OF ARTISTIC VILLA
艺墅之梦

MODERN SIMPLICITY ■ 现代简约

Located in the land of treasure with better geomantic omen in Beijing, this project is the ideal resident where celebrities can pursue artistic lives. The design is inspired by the idea of "art gallery"; the designers abandon gorgeous piles and showy designs, present artistic charms from the perspectives of details and soft decoration furniture and furnishings and convey a natural and tranquil aesthetics full of humanistic care.

The designers skillfully integrate life space and art gallery into a whole and make it grow in the home, which carries art life of the owner. Designers arrange display area, collection and communication area are set to be in the underground, whereas studio, living room and dining room are in the first floor; private rooms such as bedrooms are in the second floor. Various considerate design details bring a relaxing and quiet space feeling.

Project name: Beijing Metropolis Jiayuan B2 Show Flat
Design company: EHE DESIGN
Main materials: linen, steel, walnut, animal fur, etc.

Photographer: A Guang
Designers: Ma Jingjin, Jin Bibo, Zhu Jingru, Ge Xulian, Ma Hui

Location: Beijing
Area: 206m²

项目名称：北京名都嘉园B2样板房
设计公司：杭州易和室内设计有限公司
主要材料：麻料、钢材、胡桃木、动物皮毛等

摄影师：阿光
设计师：麻景进、金碧波、祝竞如、葛旭莲、马辉

项目地点：北京
项目面积：206m²

As for material collocation, the designers use concise and bright whitewash, stoving varnish overcoating, art bricks, partial beautiful texture white stones and black titanium steel to create a natural and leisure artistic mansion temperament which is close to life. Small ornaments of furniture are from works by European first-tier designers, which makes the space full of elegant tastes. There are no redundant ornaments; art works from the house owner are hanged on the wall, manifesting magnificent and inclusive dream of artistic villa.

White wall and sofa match with light color floors; plentiful illumination deduces elegant, plain and neat temperament of the living room. A large piece of abstract enjoyable painting and modern geometrical pattern carpet echo with each other and bring out the best in each other, making the whole space flexible. Marble background wall with ink painting texture agitates artistic feelings in everyone's heart by its magnificent natural grain. Picking up a glass of red wine to warm heart, life is so sweet.

With elegant and comfortable dining chairs, high-quality hardware and exquisite furnishings, you can enjoy good food from god and the home is full of gratitude and laughter every day.

There is no regular background wall in the master bedroom; extreme contract between black and white and avant-garde self-portrait are deeply rooted in people's mind. It is difficult to forget the once frivolous young boy. It is always remembered that those sunny and brilliant days leave many struggle stories for many years; what an aesthetic and touching time!

The owner likes simple and plain lifestyle and has an ordinary heart. Simplicity yet not simplicity and ordinary yet not ordinary are as if several pieces of paintings collected by the owner in the guest room; the painting crafts are not very excellent, but the contents can touch you deeply. Perhaps it is a treasure bought in an exhibition or a work depicted by the owner himself, these paintings are like classic movie scenes, which interprets glorious artistic life to people.

　　项目位于京城龙脉藏风聚气的风水宝地，它是名流们追求艺术生活的理想居所。设计师的创作灵感来源于"画廊"，摒弃华丽的堆砌和炫耀式的设计，从细节和软装家具饰品上呈现艺术的魅力，向人们传递一种极具人文关怀的的自然安宁美感。

　　设计师将生活空间和画廊巧妙地融为一体，让画廊生长在家中，承载主人的艺术人生。作为一个颇具艺术感的空间，设计师合理规划了空间，尽可能地满足主人的生活品味。将工作室、客厅与餐厅安排在一楼，作品展示、收藏及交流区设在负一楼，相对私密的卧室区则设在二楼，各种用心设计的细节带来轻松、安静的空间感觉。

　　在材质的搭配上，设计师运用了简洁明亮的白色涂料、烤漆饰面、艺术砖及部分有着漂亮肌理的白色石材与黑钛钢相结合，营造出自然休闲又亲近生活的艺术大宅气质。家具的小摆设都来自欧洲一线设计师的作品，让空间充满雅致的情趣。没有多余的装饰物，墙壁上挂的多是房屋主人的艺术作品，凸显大气包容的艺墅之梦。

　　白色的墙壁和沙发配上浅色的地板，更有充足的照明演绎出淡雅素净的气质客厅。大幅抽象写意画与现代几何图案的地毯遥相呼应，相得益彰，让整个空间也随之灵动起来。一抹富有水墨画笔触的大理石背景墙，以其华美的天然纹理激荡起了每个人心中的艺术情愫。随手拿起一杯红酒暖暖心，生活就是这样带有甜味。

　　优雅舒适的餐椅，还有高品质的五金件配上精美的装饰，虔诚地享用上帝赐给的美食，让家每一天都充满感恩与欢笑。

　　主卧中无常规的背景墙，大胆而鲜明的黑与白，以及硬朗前卫的自画像，无疑不植根于人们的记忆深处，难以忘怀那个曾经轻狂的青春少年。始终记得，在那些阳光灿烂的日子里，留下了许多年奋斗的故事，原来光阴也可以唯美动人。

　　喜欢清简素淡的生活方式，守一颗平凡之心。简而不简，凡而不凡，就如同主人在客房精心收藏的几幅作品，画工虽算不上特别精良，但是内容却深深打动你。也许是一次画展中淘到的宝贝，也或许是主人亲手细描的作品，这一幅幅艺术品就如同经典的电影画面向人们诠释了光影辉映的艺术人生。

LINEAR ART AND POP ART
线性艺术与波普艺术

MODERN SIMPLICITY — 现代简约

Haihang Mansion is a luxurious mansion of Haihang real estate, occupies the leading place of Daying Mount international tourism island CBD, watches the future of CBD metropolitan prosperity, is a core place with functions including government affairs, commerce, transportation and life and provides livable and beautiful life bay for middle class elites. SCD creates a delicate and exquisite, fashionable and artistic space which integrates linear art and pop art into a whole, bringing you into an avant-garde, artistic and free home. Modern furniture is concise yet not simple; cotton and linen with natural temperament collocate with leather; irregular lotus shape tea table strews at random, collocating with luxurious metal materials, casual and unrestrained, exquisite and elegant. Art reminds people of beauty and mystery and the realm of art and literature is a place with all kinds of arts. The seemingly random lines intertwine; black and white linear crossover seems irregular but pursues its own rhythm to deduce a visual feast.

Design name: Haikou Haihang Mansion North Second Area House Type A
Design company: SCD

Chief designers: Simon Chong, Canyon Xu
Soft decoration designers: Amy Du, Circe Ding

Location: Haikou, Hainan
Area: 136m²

项目名称：海口·海航豪庭北苑二区A户型
设计公司：SCD（香港）郑树芬设计事务所

主案设计师：郑树芬、徐圣凯
软装设计师：杜恒、丁静

项目地点：海南海口
项目面积：136m²

　　海航豪庭是海航地产开发的实力豪宅，独占大英山国际旅游岛 CBD 龙头之位，守望 CBD 大都会繁华未来，是集政务、商业、交通、生活等多功能为一体的核心领地，为中产精英阶级构建宜居至美生活湾区。SCD 塑造一个精致细腻，又不失时尚、艺术格调的空间，融合线性艺术与波普艺术两种艺术形式为一体，带你走进一个前卫、艺术、自由的家。具有现代感的家具，简约而不简单，带有自然气质的棉麻搭配皮革，不规则的荷叶型茶几高低错落，配合奢华的金属材质，随性不羁中却又精致优雅。艺术让人联想到美和神秘，而艺苑指艺术荟萃的地方。看似杂乱无章的线条相互缠绕，黑白线性的交叉，看似无章，却追寻自己的节奏演绎一场视觉的盛宴。

GREEN MOUNTAIN AND WHITE CLOUD

青山昼·白云栖

This project is located near Huashen Lake in Nanjing; the landscape endows it with beautiful temperament. The soft decoration integrates Oriental elements with modern materials and has other extensions; it is not pure copy or pile, but creating a traditional and charming space in modern people's aesthetic needs through the understanding and recognition of traditional culture. The designer refines symmetric and closed forms in traditional Chinese style and uses modern method to present Zen-like Chinese flavors. The collocation of walnut and cotton and linen cloth materials is suitable for the owner's pursuit of natural and plain lifestyle.

Project name: Renheng Magnolia Mountain Villa
Soft decoration: JINdesign
Main materials: imported marble, imported wallpaper, Kenai stone, walnut, oak wood, leather, cloth, etc.

Hard decoration: Nanjing Duanshun Decorative Design Consultant Co., Ltd
Photographer: ingallery™

Location: Nanjing, Jiangsu

Area: 235m²

项目名称：仁恒玉兰山庄
软装设计：JINdesign近逸设计
主要材料：进口大理石、进口壁纸、科耐石、核桃木、橡木、真皮、布艺等

硬装设计：南京端顺装饰设计顾问有限公司
摄影师：逆风笑

项目地点：江苏南京
项目面积：235m²

Layout of the living room is different from the traditional Chinese style; the designer conducts slight innovation in the closed form; side table near the sofa and two single sofas make the space more modern. The furniture cloth clarifies the relationship between black, white and gray in lightness, increasing the layering of the space. The main droplight transferred from palace lantern modeling, tea table with classical architectural charms and sofa leg wires deduce a perfect integration of Chinese flavors and modern atmosphere. The tall customized droplight in the dining room is inspired by calabash meaning "happiness and luck", presenting the owner's expectation of better life. Brass furnishings and deadwood landscaping add more fun. As master bedroom is the main rest place for the owner, comfort becomes the first consideration. No matter leather bed chair or bay window couch, all start from practicability and promote life quality.

本案位于南京花神湖畔，湖光山色的景致给予了它隽秀的气质。在软装设计中，融合了东方元素和现代材质的基础上，又另有延伸，它并非纯粹的拷贝或堆砌，而是通过对传统文化的理解和认知，以现代人的审美需求打造富有传统韵味的空间。将传统中式里对称、围合等形式进行提炼，用现代的手法来体现禅意的中式气节。核桃木加棉麻布艺的材质搭配也符合业主返璞归真的生活方式。

客厅空间不同于传统中式布局，在围合的形式上进行微创新，次沙发中的边几，以及两个单人沙发的组合都让空间更加现代。主体家具布艺也从明度上拉开了黑白灰的色彩关系，增加空间层次。宫灯造型演变过来的主吊灯，和带有古典建筑造型韵味的茶几，以及沙发角线等细节，都让空间中的中式韵味和现代气氛很好的融合。餐厅挑高的定制吊灯，造型来源于葫芦，寓意"福禄"，展现出业主对美好生活的一种期望。黄铜摆件和枯枝造景都增添了不少情趣。主卧作为主人休息的主要场所，舒适度成为第一考虑要素；不管是真皮的床尾凳，还是飘窗的躺椅都是从实用出发，再而提高生活品质。

LIFE WITH LIGHT
过有光的生活

The international trend nowadays advocates lively minimalism design; fast pace modern life needs to be simple, so does household style.

Nowadays bodies are kidnapped by reality. Tranquility and freedom are no doubt the biggest luxury in the world. We need a space to seek for spiritual tranquility and freedom. The freedom of the heart is the freedom of life from architectural space, is sunny life temperature presented by Luneng Sanya Bay Golf Villa House Type E to residents.

The open living room and dining room in the first floor use rational minimalism black, white and gray as main tones and choose masculine marble and metal, presenting a revolt against common life. A large piece of white horse art painting in the living room is a heritage of noble culture and increases cultural sense and exquisite tone of the space. The dining room is dotted with vigorous natural green, arousing the desire for sunny life. Art droplight constituted by many circles dissolves the cold and hard lines of the space. Cloth sofa and suede high back chairs soften the anxiety and all negative emotions of modern life.

Project name: Luneng Sanya Bay Golf Villa House Type E
Design company: KKD
Designer: Kenneth Ko
Photographer: A-Kuei
Location: Sanya, Hainan
Area: 278m²
Main materials: gray wood grain stone, imitated wood grain tile, technical wood, upholstery, mirror steel, etc.

项目名称：鲁能三亚湾高尔夫别墅E户型
设计公司：深圳高文安设计有限公司
设计师：高文安
摄影师：阿贵
项目地点：海南三亚
项目面积：278m²
主要材料：灰木纹石材、仿木纹瓷砖、科技木、扪皮、扪布、镜钢等

Family room is in the negative floor. Through the collocation of hard and soft decorations, the designer endows the space with more perceptual colors. Misplaced splicing log background wall is full of stereo sense with surrealistic abstract painting by American painter Robert Motherwell, manifesting art tension of modern life.

The villa has three bedrooms which perfectly extend living properties. As for the subaltern room in the basement, the designer considers the limited outside view so that he chooses red and yellow sofa decorations with fresh colors and strong visions to create livable and comfortable atmosphere to achieve life fun.

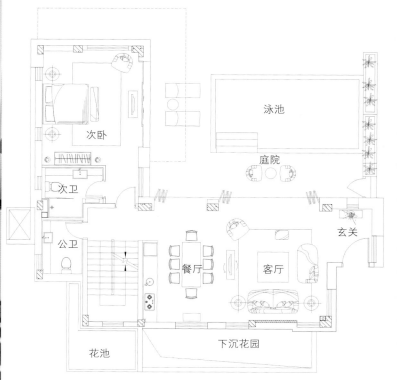

Guest room in the first floor uses light designs. White wall and natural wood veneer make the space as clean as water and respect the essence of life. Light coffee bed and cloud gray silk beddings match with cute Macarons pink short feet chairs and not fancy carpet with geometric patterns, which is simple and interesting. The master bedroom in the second floor uses lady white to set elegant tone for the space, combining with graceful gentleman black, which presents noble gesture of the owner of the villa. Velveteen chair in chic modeling near the window injects a warm color into the room, where you can have a nap. Spring scenery enters into interior through the yarn; it is the best time of life.

Outstanding interior design has always been addition first and subtraction later. Addition makes the space have a history to seek and a story to tell. Subtraction is from the designer's confidence. The simpler it is, the more powerful it is. Less sense of design leaves more sense of life to people. What Luneng Sanya Bay Golf Villa House Type E presents to us are a free life attitude of modern living, a simple, fresh and comfortable living condition and a home where you can enjoy sunshine at will.

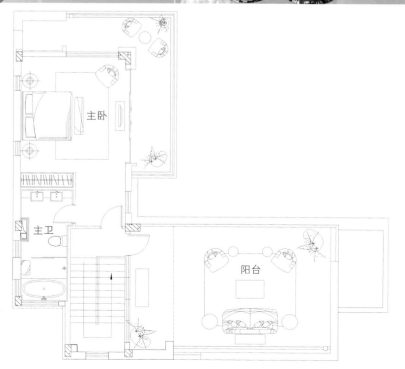
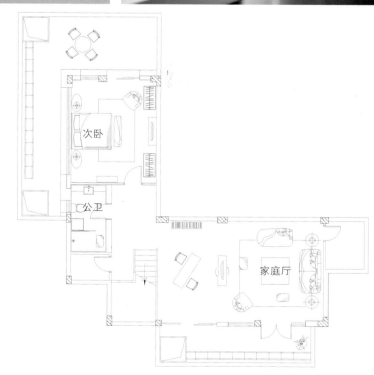

当今国际趋势，倡导明快的简约主义设计，快节奏的现代生活需要一切从简，家居风格也不例外。

在身体多被现实绑架的今天，安静和自由无疑是这个世界上最大的奢侈，我们需要一个空间，谋求心灵的安静与自由。心的自由，是建筑空间赋予生活的自由，也是鲁能三亚湾高尔夫别墅 E 户型，呈现给居住者阳光般的生活温度。

一层开放式客餐厅，以极具理性的极简黑白灰作为主色调，材质上选用阳刚的大理石与金属，体现一种对平庸生活的反叛。客厅悬挂的大幅白马艺术照，是对贵族文化的传承，增加空间的文化感与精致格调。餐厅点缀充满生命力的自然绿，唤起对阳光生活的渴望。再用多圆组成的艺术吊灯，化解空间直线条的冷硬，搭配布艺沙发、绒面高背椅，以柔软化解现代生活的焦虑以及所有的负面情绪。

家庭厅在楼下一层，设计师通过软硬装的搭配，赋予空间更多的感性色彩。错位拼接的原木背景墙，立体感十足，装饰上美国画家罗伯特·马瑟韦尔的超现实主义抽象画，彰显现实生活的艺术主张。

别墅共有三间卧室，居住属性得到完美伸张。地下层次卧，设计师考虑到窗外视野比较局限，采用了颜色鲜艳，视觉感更强的红、黄两色软装，营造宜居的舒适氛围，焕发生活的情趣。

一层客卧走的是轻设计的路线，白色墙壁、天然木饰面，仿佛清水洗涤过的空间，读懂对生活本质的尊重。浅咖色大床、云灰色丝织床上用品，配上可爱的马卡龙粉矮脚椅，并不显花哨的几何图案地毯，简单却有余味。二层主卧，以淑女白奠定空间素雅的主调，结合很有气质的绅士黑，突显别墅主人尊贵的姿态。窗边有造型别致的棉绒椅为房间注入暖心色彩，可在椅上小憩，春光林色透纱而入，正是人生最得意的时光。

优秀的室内设计，从来都是先做加法，再做减法。加法是为了让空间有历史可寻，有故事可讲。减法则是出于设计师的自信，越简单就越有力道，设计感少了，留给人的生活感就多了。鲁能三亚湾高尔夫别墅E户型，呈现的是现代家居自由的生态态度，是简单、清新、舒适的生活状态，是可以任性享受阳光的家。

UNIQUE ORIGINALITY, LEISURE RESIDENCE
独具匠心 悠然居所

This project is located in the future administrative center in Tongzhou, Beijing; the location not only edifies the temperament of learned scholars, but also avoids the hustle and bustle. No matter the overall planning or detail carving, all deduce the attitude and feeling of perfect artist. Elegant lines and modern crafts encounter in the artistic time; nobility and elegance integrate into a whole, which adds more charms. The living room uses texture painting and metal letter furnishings to collide skillfully as if playing a musical movement. Metal wine shelf in the dining room echoes with the rhythmic crystal light; fluent lines make the space stratified. In the concise yet not simple master bedroom, warm and harmonious tone reveals noble tone; the radial modeling droplight in the study echoes with the carpet, making the space stereoscopic and interesting. The overall innovative design and partial outline of different materials present the pursuit of life quality.

Project name: Vanke City Light Show Flat One
Design company: GBD
Designer: Bill Du
Photographer: Bill
Location: Beijing
Area: 130m²
Main materials: Solomon stone, beige stoving varnish plate, coffee stoving varnish plate, dark color wood floor, gray wired stainless steel, etc.

项目名称：万科城市之光样板房壹
设计公司：广州杜文彪装饰设计有限公司
设计师：杜文彪
摄影师：Bill
项目地点：北京
项目面积：130m²
主要材料：所罗门石、米色烤漆板、啡色烤漆板、深色木地板、灰色拉丝不锈钢等

本案台湖城市之光位于北京未来的行政中心通州，所处位置即熏陶了鸿儒之气又避开了丝竹喧嚣。无论从整体的规划还是细节的雕琢，都演绎着完美艺术家的态度和情调。优雅的线条与现代工匠在艺术的时光中相遇，高贵与典雅交织一体，更添韵味。客厅运用机理挂画与字母金属装饰巧妙碰撞，仿似演奏着悦耳乐章。餐厅金属酒架与律动感十足的水晶灯相呼应，流畅的线条使空间更具层次。简约而不简单的主卧，温暖和谐的色调中透着高贵格调以及书房放射式造型吊灯与地毯相呼应，使空间立体而有趣。整体独具匠心的设计与局部不同材质的勾勒，流露出对生活品质的追求。

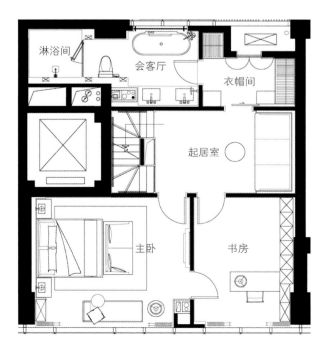

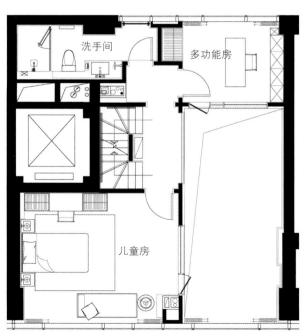

EVERY FLOOR A REALM AND SIX FLOORS SIX WORLDS
一层一境界 六层六重天

MODERN SIMPLICITY ■ 现代简约

This project is designed by the Italian designer from DOMUS who is also an impressionistic painter, boldly collocates colors, freely controls the overall situation and maturely expresses his own rich and flexible inner world. Inspired by conceptual elements of Italian top luxuries, the overall layout, material selection and furniture furnishings reflect its original and unique, fashionable and magnificent, exquisite and particular design temperament, which leaves a deep impression on people. The master of mirror reflection is proper, which enlarges the space and makes the sight neither complicated nor heavy.

Project name: Huating Villa Show Flat 13#
Design company: DOMUS
Location: Shanghai
Main materials: marble, iron art, wallpaper, etc.

项目名称：华亭首府13#别墅样板间
设计公司：上海多姆设计工程有限公司
项目地点：上海
主要材料：大理石、铁艺、壁纸等

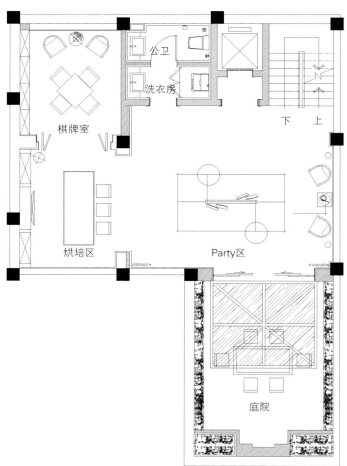

At the same time, the designer upholds people-oriented thought and customizes exclusive living spaces for each family member; every space color collocation is flexible and wonderful with different highlights. All reflects the design team's persistent pursuit of "every floor a realm and six floors six worlds" for this project.

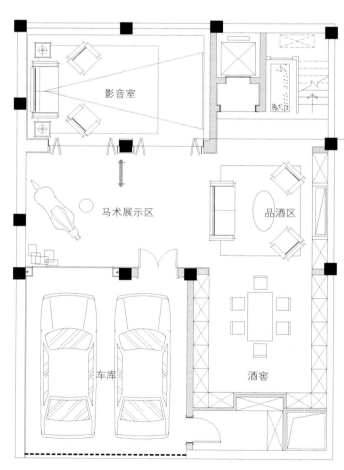

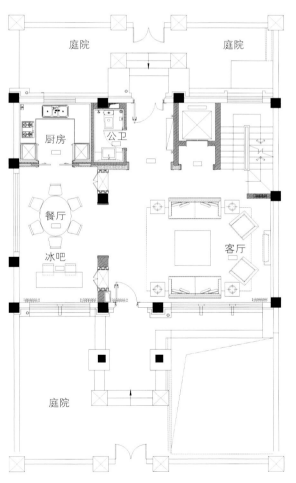
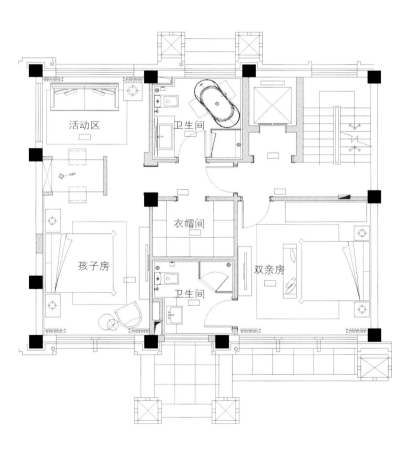

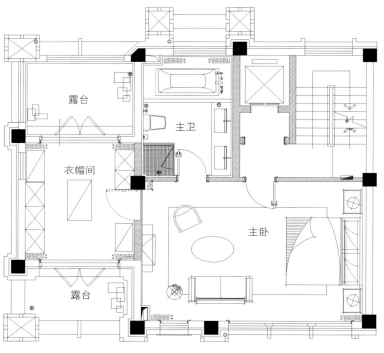

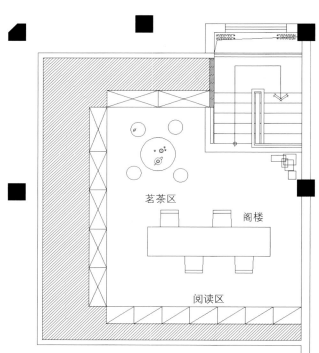

由DOMUS身兼印象派画家意大利籍设计师领衔主案，大胆地搭配色彩，自如地驾驭全局，鲜活纯熟地表达自己丰富灵动的内心世界。设计灵感来源于意大利顶级奢侈品的概念元素，从整体布局，到材质挑选，再到家具装饰，无处不体现其新颖独到、时尚大气、精致考究的设计气质。镜面反射手法的拿捏恰到好处，空间倍增的同时，视觉上却丝毫不显繁复冗赘。同时，秉承以人为本的思考，为每一位家庭成员量身定制的专属生活空间，加之每处空间配色灵动精彩，各有所显。恰恰体现了本案"一层一境界，六层六重天"设计团队的执着追求。

POSTMODE

RNISM

后现代

将现代与古典的元素完美结合起来，具有历史延续性，但又不拘泥于传统的逻辑思维方式，用理性而睿智的设计探索新的造型手法，幽默、诙谐、充满想象力。

Perfectly combining modern and classical elements, having a historical continuity, not limited by traditional logic way of thinking, exploring new modeling method through rational and wise designs, humorous, witty and full of imaginations.

TIMES WILL
时代的意志

POSTMODERNISM / 后现代

Guoxin Century Sea Garden is a first-tier river project in Small Lujiazui area near Huangpu River, is regarded as a landmark building in Lujiazui area and is a microcosm of modern Shanghai architecture. The designer starts from the unique location advantage of the project itself, looks for design inspiration and tries to create a modern and fashionable project with cultural flavors in Shanghai. So the designer positions this show flat with four bedrooms and two living rooms as modern neo-classical style.

"The highlight of the space must be the invincible river view of Huangpu River." The designer stresses that in order to bring in the river view to make it become a part of the interior decoration. They resolutely decide to change the original limited and sheltered indoor terrace into full French window. Besides, they try to open view of the entire porch from the entrance to the windowsill; without more shelters, you can see river view directly from the living room. In addition, we take steps to combine the living room with balcony and make the center of the living room closer to the river view. Because the original height of the living room is limited and gives people a depressing feeling, the designer sets several geometric mirrors in the ceiling and expands furthest the boundary of the vision to make the space look higher.

Project name: Guoxin Century Sea Garden
Design company: IADC
Main materials: light wood veneer, gray marble, bronze metal veneer, etc.
Designer: Ji Pan
Photographer: Jing Zhang
Location: Shanghai
Area: 310m²

项目名称：国信世纪海景园
设计公司：IADC涞澳设计
主要材料：浅木饰面、灰色大理石、古铜色金属饰面等
设计师：潘及
摄影师：张静
项目地点：上海
项目面积：310m²

To highlight the lifestyle of the magic city, the designer endows the entire project with features full of artistic temperament and humanistic feelings. So the entire room sets gray beige and gold as main tones, collocating with mysterious peacock blue, which forms a new, avant-garde and fashionable appearance. Gorgeous gold promotes luxurious sense of the room while peacock blue makes the space color more stratifying. The use of peacock blue is very bold, highlighting our unique understanding of area temperament of the space; using it to narrate the multicultural background in Shanghai seems to be perfect.

In addition, the designer uses natural grain of elegant gold marble to splice symmetric patterns to decorate the walls. In order to render noble temperament of the interior space, the designer uses Art Deco and modern style to create main tone of the whole space design; especially the anaglyptic customized background wall with the Bund scenery becomes the highlight of the entire room. The architectural skyline near Huangpu River is extended to the interior; the interior and exterior echo with each other and integrate into a whole; the image of the entire city is condensed into a focus.

The special customized Swarovski frameless crystal droplight presents the nobility of the space. The diamond modeling lamp furnishings skillfully combine with the entire hard base, which strengthens visual transparent sense of the space and presents unique taste and temperament of the elites in this era. This punchline brings the river view into interior; the scenery and furniture modeling of the whole space coordinate with each other.

On space materials, the designer chooses precious velvet and special leather to match with unique texture brought by cooper, endowing the whole space with calm and restrained features. Every furnishing indoor is picked carefully by the designer; every texture has exquisite tactile impression, presenting delicate taste owned by Shanghai.

Space and life are in an interactive relation and existence; interior design cannot exist independently without environment. So when we finish the entire project, everything seems to be natural. A unique individualized space is born; it is not only the product of design, but also the reflection of times will.

国信世纪海景园是小陆家嘴地区黄浦江畔的一线江景项目，称得上是陆家嘴地区标志性的建筑，也是现代上海的建筑缩影。设计师从项目自身独特的地理位置优势出发，寻找设计的灵感，试图构建一个充满现代时尚又具有上海特有文化气息的作品。因此，设计师将这套四室两厅的样板房，定位为现代新古典风格的设计。

"空间的亮点必定是黄浦江的无敌江景。"设计师强调，为了引入江景，令其成为室内装饰的一部分，我们毅然决定将原本受局限与遮挡的室内阳台，改为全落地窗的形式。从入口到窗台的距离，我们力图打开整个玄关的视野，不做更多的遮挡，你可以通过客厅直接看到江景。此外，我们采取将客厅与阳台结合起来整体考虑的思路，让客厅的中心点尽可能地更靠近江景。由于客厅原本的层高有限，会让人有压抑的感觉，设计师在天花上设置了大负面的几何镜面，最大限度地拓展了视觉的边界，使空间看上去更高。

为了凸显魔力之都的生活形式，设计师为整个项目设定了具有艺术气质与人文情怀强烈感受的特质。因此，整个房间以米灰色与金色为主色调，配以神秘的孔雀蓝为搭配，形成新锐、前卫、时尚的面貌。华丽的金色点缀，提升了房间的奢华之感，而孔雀蓝则令空间在色彩上更具层次。孔雀蓝的使用非常大胆，凸显了我们对于项目所在地域气质的独特理解，用它讲述上海的多元文化背景似乎再合适不过。

此外，设计师还利用雅致金大理石的天然纹理，拼合出对称极强的拼花来装饰墙面。为了渲染室内空间的矜贵气质，设计师运用了 Art Deco 与现代风格相互融合手法来构建整个空间设计的主基调，尤其是浮雕感的定制外滩江景背景墙，成为整个房间的亮点，黄浦江边的建筑天际线被延伸至室内，室内与室外遥相呼应，融为一体，这座城市整体的形象被浓缩成了焦点。

特意定制的施华洛世奇无骨架水晶吊灯，尽显空间的尊贵。钻石造型的灯饰构件，巧妙地与整体硬座结合，既增强了空间视觉的通透感，又显露出属于这个时代精英阶层的独特品味与气质。这一点睛之笔，使得引进江景入室内的同时，景致与整体空间内家具的造型，也互相协调了起来。

在空间取材上，设计师选用了珍贵的丝绒、特殊的皮革，来搭配铜所独具的质感，赋予整个空间沉稳、内敛的特质。居室内的每一件摆设都经由设计师用心挑选，点滴的质感拥有细腻的触摸感受，呈现出上海贯有的精致品味。

空间与人生一种交流互动的关系和存在，室内设计决不能剥离于环境而独自存在，因此，当我们把整个作品完成时，一切都似乎是水到渠成的——一个独特的个性空间就此诞生，它不仅是设计的产物，也是这个时代意志的体现。

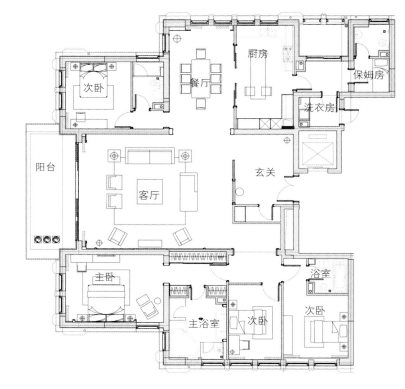

SCENORGRAPHY INTERIOR
阁楼·戏剧

The loft is located in the commercial center, Zhongzhi Xiangteng City Square, which is called "Nanxiang international community" and "the first CBD Nanxiang places of wise in Shanghai suburb". In this LOFT space with a design area of about 110 square meters, the designer integrates Chinese and Western elements, uses stage design techniques, creates different scenarios and skillfully fuses them together. Light and shadow interlace and the colors are rich and warm, creating a leisure, tranquil and comfortable living atmosphere.

Fresh black and white of the porch complement each other; the half-open red curtain uncovers the mysteries of the small room. To present an open living room, the designer specially uses transparent yellow and blue filmed glasses on the bar which is full of heavy leisure, fashionable and modern breath under the foil of black and white floors.

The designer uses a two-meter-wide concise partition wall which divides the narrow space into living room and study. So one viewer smiles and says "though the space is small, it has everything". One side is strong and rich while the other side is fresh and elegant. The partition wall forms a strong visual contrast as if the fresh and jumping change of stage scene.

Design company: Ivan C. Design Limited
Designer: Ivan Cheng
Main materials: Blue Shanghai White, Hutton, Fullart, etc.

Photographer: Wei Ji
Location: Shanghai

Area: 110m²

设计公司：Ivan C. Design Limited
设计师：郑仕樑
主要材料：海上青花、赫顿、富雅等

摄影师：恽伟
项目地点：上海

项目面积：110 m²

The tine living room uses mirror to maximize sense of extension. Mirror goldfish stickers in the ceiling and blue carpet on the ground form a corresponding echo of blue sea and jumping fish. Black and white parrots fly all day and stop to perch here for a while. Warm dragonfly lights integrate man and nature in life. Fashion and bright red KARTELL chairs and Chinese drum stools echo with the Chinese dresser at the corner, which are linked together by three red lines in the carpet elaborately designed by the designer. Different textures are collided here, forming a complete color collocation.

Entering the study, there are concise and decent desk, customized walnut wood chest of drawers and Chinese white bookcase in the study which is retro and modern and presents calmness and restraint by its unique gesture. The dark coffee armchair originated by the designer has stereo back modeling. The designer skillfully integrates Chinese hardware with white bookcase between which there are pottery furnishings bought from different places.

Combining with the arc architectural structure, the designer uses s-shaped interlayer margin to create chic tall space. The designer pursues traces of his previous years, combines gray lens with bright mirror, uses arc modeling and matches Roman chapter and enlarged abstract external window modeling with towering sculpture in front of the mirror. Sculpture concept derives from *The Endless Column* by Romania sculptor Constantin Brancusi, which means the pillar from the ground to the sky that pushes life energy into the endless space is the support which makes the sky link with the ground. Lights through blue, yellow sticker glass show a classical and lively church space atmosphere.

This church is naturally accompanied by angels. There is *Rainbow Angel* by famous Chinese contemporary sculptor Qu Guangci on the table, which is different from western angel and is the angel owned by Chinese which outlooks the future of Chinese themselves. Under the windowsill, there are a bench specially designed by the designer, a round blue and white porcelain table originated by the designer which is completed by blue and white porcelain veneers with classical modeling. A tall Chinese official cap chair, a rotatable Blue Shanghai White brand dining chair and a portrait of Chairman Mao by Andy Warhol match with Chinese and English black chair made by the designer. Its back is inlaid with red cheongsam modeling ceramic with a stereo back modeling, interpreting the historical changes of China since ancient times. The elephant accessory on the windowsill adds a sense of calmness and is full of aura. It presents an interior outside scenery and a Chinese and Western courtyard.

Looking around in the "yard", there is a pair of "never separating" sculpture downstairs and an unintentionally designed sculpture, that is the winding stair without handrails. S-shaped margin of the second floor is like a floating pavilion in the air. Upstairs, there is *Frenetic Wing* by a young contemporary artist Ben Gough on the wall as if the flying bird flutters and soars high and the free pigeon brings peace.

Passing through the white frame in the second floor is like the change of stage scene, then you immediately turn to the bedroom and enter into a "In the Mood for Love" tone. A Chinese old double chair bought by the designer reflects the entire bedroom. The bed background wall is a right-left symmetrical pattern with purple and black wallpaper in the center. The two sides are bright mirror and black and white chest curtains which take place of cupboard doors to balance the hale space. The left is a dragonfly desk lamp while the right is an ultramodern hat droplight; lights from the two lamps are dark and dim. Two piece of concise and retro furniture in Shanghai style at the bedside and an old small fan tell a love story as if in the old and tranquil Shanghai lane under the weak lights.

The front of the bed presents a different scene with gorgeous colors. A piece of poster of the film *In the Mood for Love*, a lazy double sofa chair, a scarf casually hung on the Chinese hanger, an abstract oil painting *Blue 1984* by Hong Kong artist Mr Xu Yuka, the round carpet and a transparent blue floor glass form a comedy effect, tranquil and leisure, as if a reproduce of the scene in the film: "Li-Chen Sun" sits here, lights a cigarette, doesn't smoke it here and puts it to let it fumy, hovering and scattering; the scene is delicate and gentle. Through the eight prism on the wall, entering the shuttling time, swaying cheongsam and dim lights and so on with dancing butterflies

This is a kind of feeling to old Hong Kong and a kind of recall of blossom age!

阁楼地处誉为"南翔国际居住社区"和"上海市郊第一CBD南翔智地"的商业中心——中冶祥腾城市广场内。在设计面积约110平方米的LOFT空间内，设计师融贯中西元素，运用舞台设计手法，创作不同的场景，却又巧妙地融合在一起，光影交错，色彩丰富温馨，凝造悠闲、静谧、舒适的居所氛围。

玄关鲜明的黑白色调相得益彰，在半掩的红色帷帐下，揭开斗室的神秘面纱。为了呈现开放式的客厅空间，设计师特意将吧台用通透的黄、蓝色贴膜玻璃拼接搭成，在黑白格地砖映衬下，富有浓郁的悠闲时尚现代气息。

设计师利用2米宽的简易隔墙，将狭小的空间分割出客厅和书房区域。于是有观者笑言"空间虽小，内有乾坤"。一边浓烈丰富，一边清新典雅。一墙之隔，形成强烈的视觉对比冲击，犹如舞台转景般鲜明跳跃着。

在客厅微小的空间，更通过镜子，得到最大化的延伸。天花镜面金鱼贴画，和地面蓝色地毯，形成上下呼应的海蓝鱼跃画面。黑白相间中的鹦鹉，昼夜翱翔，纷纷驻足栖息于此。再点缀温馨的蜻蜓灯光，将生活其中的人与自然巧妙地融合。时尚通明的红色KARTELL椅和中式鼓凳，与拐角处的中式梳妆台遥相呼应，并通过设计师精心设计的地毯中三条红线连接在一起，不同的肌理在此碰撞，形成完整的色系搭配。

步入书房，有简洁大方的办公桌，特别定制的胡桃木质斗柜和中式白色书柜，几分复古，几分现代，以它独有的姿态彰显出些许沉稳而内敛。其间一把设计师原创的深咖扶手椅，拥有立体的后背造型。设计师将中式五金件巧妙地与白色书柜糅合，中间摆放着设计师从不同地方淘来的陶艺等饰品。

　　结合弧形的建筑格局，设计师采用S形隔层收边，营造别致的挑高空间。设计师追逐着自己早年的印迹，将灰镜与明镜的结合，弧线造型，罗马柱头的运用，和放大的外窗抽象造型，搭配镜前高耸的雕塑。雕塑概念源自罗马尼亚雕塑家康斯坦丁·布兰诺西(Constantin Brancusi)的作品《无尽之柱》(the Endless Column)，寓意：从地上向天的支柱，将生命的能量推向无尽的空间，是支撑天空连接天地的支柱。透过蓝、黄色贴膜玻璃的光线，呈现出古典而明快的教堂空间氛围。

　　教堂自然有天使相伴。桌上摆放着中国当代著名雕塑家瞿广慈的作品《彩虹天使》，不同于西方的天使，这是中国人自己的天使，展望着中国人自己的未来。窗台下是设计师特别定制的长椅，一款设计师原创的青花瓷圆桌，以古典造型，青花瓷饰面完成。一张高高的中式官帽椅，和可转动的海上青花品牌餐椅，一幅安迪·沃霍尔(Andy Warhol)的毛泽东肖像画，搭配设计师自己设计的中英混搭黑色椅子，背部镶嵌红色旗袍造型陶瓷，后背立体造型，诠释着中国从古至今的历史变迁。窗台上的大象饰品，平添了几分稳重而富有灵性。整体呈现的是室内的户外风光，一个中西融合的院落。

　　站在"小院"环顾，楼梯下是一对"永不分离"的雕塑，还有设计中无意的雕塑——没有扶手的楼梯曲折向上。2F隔层S形收边，仿佛漂浮空中的楼阁。沿楼梯拾级而上，墙壁上是当代青年艺术家本·高夫(Ben Gough)的作品《狂热翼》(Frenetic Wing)，如翱翔的鸟儿，振翅高飞；如自由的鸽子，带来平和。

　　穿过2F的白色门框,如同舞台布景转换一样,立刻转入卧室空间,进入"花样年华"的格调中。一款设计师淘来的中国老式双人椅,透视着整个卧室空间。床背一面呈左右对称的格局,中间紫黑相间的壁纸,两侧是明镜和黑白色衣柜幕帘(以布帘代替柜门,平衡了硬朗的空间),左侧一盏蜻蜓台灯,右侧一盏超现实的帽子吊灯两盏灯光照射下,昏暗而朦胧。床头两个简易怀旧的海派家具,一台老旧的小风扇,在屏弱的灯光下,仿佛来到60年代老上海幽静的弄堂,诉说着一段欲说还休的情感故事。

　　床的正面则呈现另外一番景象,色彩绚丽多姿,五彩缤纷。一幅电影《花样年华》海报挂画,慵懒的双人沙发椅,一条围巾随意搭在中式衣架上,在抽象的油画,香港艺术家许尤嘉先生的作品《蓝1984》和圆形地毯背景映射下,加之通透的蓝色地面玻璃,颇具喜剧效果,宁静而闲适,仿佛影片中的某个镜头在这里重现:"孙丽珍"呆坐在那里,点燃了一支香烟,不抽,只是放在那里任它的烟雾缭绕、盘旋、四散,画面细腻温婉。透过墙壁上的八棱镜,随着翩翩起舞的蝴蝶走进穿越的时光,摇曳的旗袍,昏暗的灯光……

　　这是对旧香港的一种情怀,是对花样年华的一种追忆!

POSTMODERNISM 后现代

NOBLE · CLOWN
贵族·小丑

The owner wants to find a peace land in the busy Hangzhou city to have a rest and enjoy the drama life. Immersed in the drama stage designed by himself, the designer deduces a relaxing and humorous pleasant space which inadvertently reveals a noble flavor through comedy clown. The gray tone is injected with blue, yellow and purple red bright colors; the skillful and heavy colors create a relaxing, lively and leisure slow pace life space for the owner to percept deposits of years in life.

Entering the door, the lively song *Swan Lake* by famous Chinese contemporary sculptor Qu Guangci welcomes every guest's arrival. Customized striking rainbow cabinet is their gorgeous stage, Africa portraits with chrysanthemum headwear are their musicians and the beautiful vase girls at side become their dancers. A witty and humorous drama opens this stage play.

Design project: Hangzhou Sky Lake Wenxi Celebrity Garden Building 11 No. 506
Design company: Ivan C. Design Limited
Main materials: Fullart, Hanshan Tang, Fannou, etc.
Designer: Ivan Cheng
Photographers: Wei Ji, Kang Zhi
Location: Hangzhou, Zhejiang
Area: 85m²

项目名称：杭州天湖文溪鼎园11幢506
设计公司：Ivan C. Design Limited
主要材料：富雅、涵山堂、法诺等
设计师：郑仕樑
摄影师：恽伟、支抗
项目地点：浙江杭州
项目面积：85m²

It is not only the interpretation of the designer's life but also the intersection with other people's life. Specially arranged salon area presents the ancient aristocratic banquet scene. The designer designs a 3.6 meters long table with colorful small squares on the surface, deducing the exquisiteness and art of noble decoration. A European lace series chair *The Secret of Duke William* is put on the end as if a duke presides over a magnificent dinner. Different "guests" sit around the table and enjoy good wine and dainty dishes in the banquet. Under the foil of dreamy lights, there seems to be many attendants in the ground mirror broadside who are elegantly severing their duke and duchess. There are cloisonne vases with Chinese feature whose decorative patterns are the common red leaves in local Hangzhou and many ceramic artworks on the nearby display cabinet; the combination of Chinese and Western culture is distinctive. The removable wine cabinet on the other side with slightly retro leather suitcase elements foils a sense of reminiscence in steam train era as if on the scene. The opposite continues the atmosphere of a stage play; "musicians" deduce classic classical music with people. Behind the Chinese screen, there is a different scene. Fan-shaped armchair with some amorous feelings in old Shanghai, coffee table in different shape, leather sofa in royal chair shape, French side cabinet and blue red stone bench make you associate that in the old Shanghai villa, several dressed up young men and women are drinking coffee and chatting cheerfully, manifesting the petty bourgeoisie taste. The large piece of yellow green background wall and red and green tone of the space continue the noble color in salon area, deducing an elegant play in old Shanghai style.

Walking into the bedroom, it is another kind of tranquility and fun. Blue space is interspersed with gold; velvet curtains and beddings match with clown painting inspired by the work *Mr. Blue Concerto* from young artist Lv Yanxiang; the enjoyment of five senses makes you feel sleepy immediately. The dramatic creation endows the space with a dreamlike scenario. The clown "Mr. Blue" in the Baroque frame acts as a guard while Chinese arhat bed, natural brown *The Islander Fan*, *Captain S Desk Chair* and blue abstract painting by Canadian artist seem to be in dreams of the owner, flying freely...

This is not the owner's home, yet it presents previous traces of the owner; this is not the owner's permanent home, yet it deduces stage segments of the owner's drama life; I, am the "noble"; I, am the "clown".

　　于杭城闹市找寻一隅安静的乐土，偷得半日闲，品味如戏的人生。沉浸在自己设计的戏剧舞台中，设计师以喜剧小丑演绎一个轻松诙谐的欢愉空间，却在不经意间流露出贵族的气息。于灰调中融入蓝色、黄色、紫红色等明丽色块，巧妙而浓郁的色彩调和，营造轻松、活泼、闲适的慢生活空间，感知人生岁月的积淀。

　　一进门，一曲欢乐的《天鹅湖》（中国当代著名雕塑家瞿广慈作品）迎接着每一位客人的到来。私人订制的醒目彩虹柜是他们绚丽的舞台，菊花头饰的非洲人像则成了他们的乐师，旁边还有美丽的花瓶姑娘为他们伴舞。诙谐、幽默的戏剧画面拉开了这幕舞台剧。

　　这里演绎的不仅仅是一个设计师的人生，同时诠释了与其他人人生的交集。特别布置的沙龙区，呈现了古代贵族式的宴会场景。设计师亲自设计的3.6米长桌，桌面上拼接着五彩的小方砖，演绎了贵族装饰的精美与艺术。一款欧式蕾丝系列的座椅《威廉公爵的秘密》放于一端，犹如一位公爵主持着一场华丽的晚宴。形色各异的"客人"围绕着长桌，享用着晚宴带来的美酒佳肴。在如梦似幻的灯光映衬下，侧边的落地平镜中仿佛站着很多随从，为他们的公爵和夫人优雅的服务着。旁边的陈列柜中，珍藏着富有中国特色的景泰蓝花瓶，花纹为杭州当地常见的红叶，还有很多的陶瓷艺术品，中西文化的结合别有风味。另一侧略带复古皮箱元素的移动酒柜更烘托了几分蒸汽火车时代的怀旧感，如临其境。对面则延续了舞台剧的氛围，"乐师"与大家演绎着经典的古典音乐。

　　隔着中式屏风，则是另一幕剧。窗边带着些许老上海风情的扇形靠椅、异形的咖啡桌、皮质贵妃椅式沙发、法式边柜、蓝红色的石凳，让人浮想起老上海的洋房内，几位打扮精致的青年男女喝着咖啡，谈笑风生，小资情调尽显无遗。而墙身大幅的黄绿色背景，和空间红中带绿的格调延续了沙龙区的贵族色彩，演绎的是一幕老上海式的优雅剧。

　　隔着中式屏风，则是另一幕剧。窗边带着些许老上海风情的扇形靠椅、异形的咖啡桌、皮质贵妃椅式沙发、法式边柜、蓝红色的石凳，让人浮想起老上海的洋房内，几位打扮精致的青年男女喝着咖啡，谈笑风生，小资情调尽显无遗。而墙身大幅的黄绿色背景，和空间红中带绿的格调延续了沙龙区的贵族色彩，演绎的是一幕老上海式的优雅剧。

　　这并非主人的住所，却不时呈现主人曾经的痕迹；这不是主人常住之地，却演绎着主人戏剧人生的舞台片段：我，就是那个"贵族"；我，就是那个"小丑"。

TONALITY
格调

The controversial points of the design are that visual exhaustion created by "no order" piles of different colors, acosmia feelings caused by wrong textures between folk custom native cloth and formal dress and Eastern and Western cultural gap between them. Hence the designer says: "contrast creates conflict; it is a double-edged sword, no matter the hard collisions among the same purity, lightness and toughness or forcibly welding of different genres, creating visual noises can activate the space. The premise of balance is to fill up the gap in the center to make both sides harmonious." These are presented in the space design. The whole space uses stick figure method to outline the abstract frames and uses borrowing scenery to render; inside and outside of the painting echo with each other. The corridor white space dilute the aftershocks of conflict between red and green; the water is easy to overflow when it is full; the eclipse is easy to happen when the moon is full; blocking is not as good as dredging. Red, green, black, white and yellow are used repeatedly and throughout the overall design, bringing visual shocks and presenting the designer's design concept which is not limited by traditional one.

Design company: Silk And Satins Interior Design Studio
Designer: Junman Lian / www.tofree.com.cn
Main materials: tile, wallpaper, northeast China ash panel, latex paint, solid wood floor, etc.

Photographer: Yuedong Zhou
Location: Fuzhou, Fujian

Aera: 140m²

设计公司：云想衣裳室内设计工作室
设计师：连君曼 www.tofree.com.cn
主要材料：瓷砖、墙纸、水曲柳面板、乳胶漆、实木地板等

摄影师：周跃东
项目地点：福建福州

项目面积：140m²

设计的"争议点"在于不同花色的"无主次"叠加引发视觉疲惫,民俗土布与礼服之间质感错位的违和感,以及横亘其中的东西方文化沟壑。设计师对此有话要说:"对比制造冲突,属于双刃剑,不管是相同纯度、明度、面积的硬碰硬,还是不同流派的强行焊接,制造视觉噪音的同时也可能是激活空间的兴奋剂。画面平衡的前提是填补上中间落差的节拍,使矛盾双方的对立不至于图穷匕见面面相觑。"这些都体现在空间设计中:整个空间用简笔画的手法勾勒出抽象的骨架,利用借景重彩渲染,画里画外,相映成趣;廊道留白稀释了红绿对抗的余震,水满则溢,月盈则亏,堵不如疏;红绿黑白黄循环重复,贯穿整套设计,带来视觉震撼感的同时,表现设计师不宥于传统的设计理念。

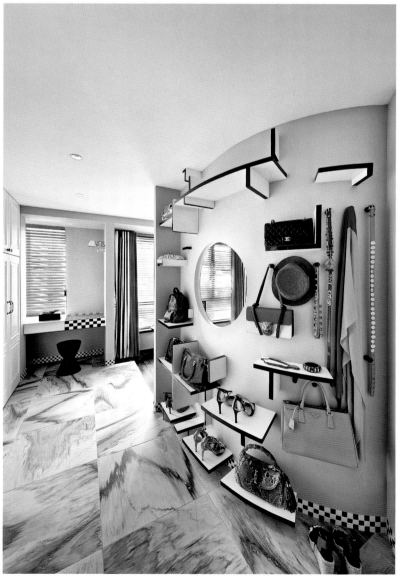

ONE DAY IN PARIS
巴黎浮生

This time Dariel Studio returns to his hometown Paris. In the central area which is nourished by art and culture, namely, 4e arrondissement de Paris, he creates the first private apartment project of his studio in France.

As the founder and chief designer of the studio, Thomas Dariel adds his personal frolicking surreal footnotes on the basis of Paris's elegant and exquisite tone. He freely develops imagination and creativity in the blank space and takes the owner's expectation into consideration which is to own a personalized place in Paris with profound culture and connotation.

Entering the apartment, it is amazing; the viewers seem to enter into a deep blue space. Wall, ceiling and blank door are treated with monochromatic color and all surfaces are integrated and intersected. The space with strong colors guides the viewers to the bright place, that is the entrance of the living room.

Project name: Private Apartment in Mulholland Avenue, 4e arrondissement de Paris
Design company: Dariel Studio
Designer: Thomas Dariel
Location: Paris
Area: 140 m²

项目名称：巴黎四区穆赫兰大道私人公寓
设计公司：Dariel Studio
设计师：Thomas Dariel
项目地点：巴黎
项目面积：140 m²

The corridor floor is covered with customized geometric carpet which not only supports the structure of the space, but also brings gentle warmth. This six-meter long carpet continues to the entrance of the bedroom and forms an entire long corridor. At the end of the corridor, there is a celadon porch table designed by the designer; this piece of poetic article skillfully softens the corner.

The living room continues the designer's distinctive style which is bright and pleasant and full of various lines and color collisions. The designer is good at adding the sense of humor featured by French into elegant and complicated interior design. The super size printed oil painting *Portalen Portalen* from the artist Kay Sage uses its surreal vision to set a tone for the space.

Wall of the living room is horizontally divided into two colors and turns in different facades. The geometric pattern of the TV cabinet constituted by log and black color, contrast color of Lazy Susan tea table from Maison DADA, these kind of color philosophy is continued in the furniture. Gray pink side cabinet breaks the routine to use "umbrella handle" single foot support; India Madhavi designs the lamps with petals; little Eliah lamp type droplight in the air challenges the law of gravity, and so on. The designer uses these unusual articles to hint the playboy style in 1930s.

Hermes sofa, Se London dining chairs, Tai Ping carpet and new brand Maison La Chance side table present the preference of the owner and designer to unique furniture.

The bathroom is the epitome of the entire apartment. The designer abandons traditional pure white marbles and uses wallpaper with complicated floral patterns to bring the bathroom an intoxicating smell of tropical rainforest.

From ordinary to extraordinary, it is rigorous and nifty with freedom from inside to outside.

　　这一次 Dariel Studio 回到了故乡巴黎，在这片被艺术文化滋养的土地的中心位置——巴黎四区穆赫兰大道，实现了事务所在法国的第一个私人公寓项目。

　　作为事务所的创始人和首席设计师，Thomas Dariel 在巴黎优雅精致的基调上，添加了他个人嬉闹玩乐的超现实注脚。在这个空白空间中自由的发挥想象力、创造力，同时又兼顾了业主的期待：在文化底蕴深厚的巴黎拥有一片属于自己的个性领地。

　　一入公寓即令人惊叹，观者仿佛陷入一个深邃的蓝色空间。墙壁、天花板、暗门都统一作单色处理，所有表面的界限融化交错。这一强烈的色彩空间引导着观者的脚步直到光线明亮之处——起居室的入口。

　　走廊地板上是定制的几何图形地毯，不仅支撑着空间结构，同时带来柔和的暖意。这条长约 6 米的地毯将一直贯穿至卧室入口，形成一条完整的长廊过道。在走廊的尽头是设计师设计的青瓷色玄关桌，这件充满诗意的物件巧妙柔化了转角。

　　起居室同样延续着设计师的鲜明风格，明亮又欢愉，充满各种线条与颜色冲撞。设计师擅长在优雅复杂的室内设计中增加法国人特有的幽默感。来自艺术家 Kay Sage 的超大尺寸《Portalen Portalen》印刷油画用超现实视角为空间定调。

  起居室墙面被水平分割为两种色彩，并在不同的立面反转。电视柜原木与黑色构成的几何图形，Maison DADA 的 Lazy Susan 茶几的撞色，这种色彩哲学在家具中延续着。灰粉色边柜打破常规，使用"伞柄"单脚支撑；由 India Madhavi 设计台灯自带脚踏；悬浮在空中的 little Eliah 台灯式吊灯挑战重力定律……设计师巧用这些超脱常规的物件暗示了 20 世纪 30 年代的花花公子格调。

  Hermes 的沙发，Se London 的餐椅，太平地毯以及新晋品牌 Maison La Chance 的边桌，体现了业主和设计师对独特家具的偏好。

  浴室是整个公寓的集大成者，设计师抛弃了传统的纯白色大理石材料，改用绘有繁复花朵图案的墙纸，为浴室空间带来了热带雨林的醉人气息。

  从平凡到非凡，严谨中不失俏皮，由内而外的自在。

METROPOLIS AND YUPPIE
都市·雅痞

POSTMODERNISM 后现代

He is the first fashion guest during the day who cannot survive without madness; at night, he becomes an innocent and evil dissolute gentleman who puts passion in parties and releases all lights.

Yuppie needs indulgence, needs unrestraint and makes no secrets of his likes and dislikes. Loving Harley is a kind of faith; the resonance of heart and motor make hormone grow madly; the roar of rock music can mobilize nerves and drive the body as if the delight of ice burnt facing the wind. Time teaches me to do happy things; this is the character of a gentleman.

Time teaches me to do happy things; this is the character of a gentleman.

Project name: Nanchang Greenbelt Chaoyang Center Creative House Type Show Flat
Space design: SLD
Soft decoration: Imaging Design
Location: Nanchang, Jiangxi

项目名称：南昌绿地朝阳中心创意户型样板间
空间设计：梁志天设计师有限公司（SLD）
软装设计：成象软装
项目地点：江西南昌

Give yourself a direction, you can see sunsets; with a glass of whisky, you can have a short encounter with the slight drunk and return to the romance of a unfettered man. Wood under the light is like the color near the sky, which has fun between light and shadow. Warm or wild, cute or happy, every small funny furnishing in the house can tell a story. Poking fun at himself is a small taste of life; he captures the trend of life. Holding a playing attitude, every corner is an "unserious" fashion class.

Luxurious at will, a close distance is separated by mountains and seas. Water and ink flow, which is jumping and exaggerating as if the unrestrained life. Elegant plates as if from food artist are the sentimental everyday of the hostess. The encounter of pure color and feeling is cool, beautiful and romantic. The poetic light emits a love of aesthetics; wake up, you are in the opposite bank. Warm lights sparkle in the bed, emitting warm texture; the gripping string fades away along with the sun.

　　白天是时尚的首发客,不疯魔不成活,夜晚降临,他变成天真有邪的浪荡绅士,将热情糅合在派对里,释放所有光芒。

　　需要放纵,需要洒脱,雅皮士毫不掩饰自己的爱恶。热爱哈雷是一种信仰,心脏和马达的共鸣让荷尔蒙狂野生长,听呐喊嘶吼的摇滚乐,调动神经,驱动身体,似冰迎风燃烧的痛快。

　　时间它教会了我,只做开心事,这,是绅士的品格。

给自己一个方向，可以看到日落日升的模样，一杯威士忌，便与微醺短暂的偶遇，回归逍遥客的浪漫。光晕下的木质的似天边微微泛起的色彩，浮光掠影间玩味随性。或温暖或狂野，可呆萌可欢脱，家中每件有趣的小摆设都可讲出一段故事。调侃自己是对生活的小情趣，他是捕捉潮流的生活家。抱着玩乐的姿态，每一个角落都是"不正经"的时髦课。

随性试戴浮华，咫尺之间，隔山海。水走墨流，跳脱张狂，似对生活的潇洒不羁。美食艺术家般的优雅摆盘，是女主人有情调的日常。纯色与情怀的邂逅，冷艳却带着浪漫。光的诗意弥散点染出对美学的迷恋，醒来，你已在彼岸。暖光洒在床头散发温馨的质感，波动心扉的弦随日光逐渐消散。

EXTREME DEDUCTION

极致演绎

POSTMODERNISM ■ 后现代

Almost all great designers will say: "My design inspiration comes from life." If this sentence is extended, it must be "exquisite and elegant life". This show flat exactly interprets this kind of exquisiteness and elegance.
The overall space layout fully considers the feature of long width of townhouse and uses compact and full rhythms to decorate. Blue and green which shine glories in the fashion city Milan are used throughout the entire space, which is quiet and comfortable. A poet said that "music makes life more elegant". White piano has occupied small half of the living room. This home must highlight life quality to make life full of elegant and happy flavors.

Project name: Shanghai Xinhong Villa Garden Villa Show Flat Townhouse Type
Design company: La More Decorative Design Co., Ltd
Main materials: marble, leather, cloth, etc.

Location: Shanghai
Area: 300 m²

项目名称：上海新弘墅园别墅样板房·联排户型
设计公司：乐摩装饰设计（上海）有限公司
主要材料：大理石、皮革、布艺等

项目地点：上海
项目面积：300 m²

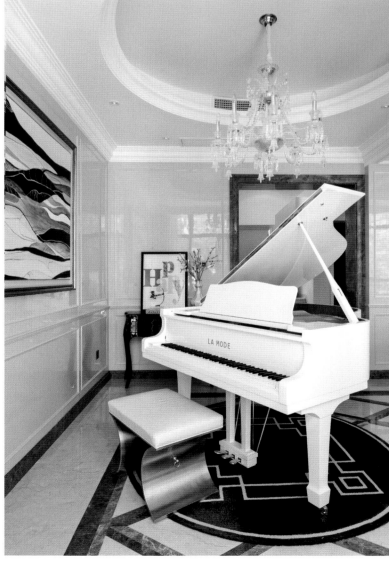

The host is a fan of polo which was called "King Movement" when born. The basement is a place for the owner and his ball friends to choose equipments, rest and chat. All equipments, souvenirs and honorable trophies about polo are presented here. Chinese claborate-style painting, Western realistic painting and modern abstract painting about polo on the wall present the owner's love to polo and deduce a polo cultural picture scroll traversing through ancient, modern, Oriental and Western times. A special made wine cellar collects hundreds of bottles of wine, waiting the back of kings who will open the fragrance and toast with each other under the shadow of cups.

Lights of the master bedroom are soft, collocating with light purple and clean white, which makes this space show more temperament and taste of the hostess. Private master bathroom and tub near the window in the master bedroom add a little romance for the space. Bright blue matches with charming dark coffee in the subaltern room, which is clean and cool; extreme and comfortable bed matches with full soft coverage walls, making all guests feel at home.

几乎所有伟大的设计师都会说："我设计的灵感源自生活"。如果将这句话延伸，那一定是"精致而优雅的生活"，本套样板房就为大家诠释了这份精致与优雅。

在整个空间布局上，充分考虑到联排超长开间的特点，以紧凑而饱满的节奏进行装点，在时尚之都米兰大放光彩的蓝绿色，贯穿整个空间，安静、惬意。有诗人说"音乐让生活更优雅"，白色的钢琴就占了小半个客厅，可见这是一个极其注重生活品质的家，让家中时刻洋溢着优雅与幸福的味道。

男主人是一个马球爱好者，马球自诞生起就被赞誉为"王者运动"，地下室是主人与球友们挑选装备、休憩畅谈的场所——与马球相关的装备、纪念品、见证着荣誉的奖杯，都在这里一一展现。墙上关于马球的中式工笔画、西式写实画、现代抽象画，既体现了男主人对马球的热爱，也交织演绎出一幅穿越古今东西的马球文化画卷。一个特别定制的酒窖，珍藏了上百瓶佳酿，随时恭候王者们运动归来，开启芬芳，觥筹交错，杯影摇曳。

主卧线条柔和，搭配淡淡的紫，干净的白，这个空间更多透露出女主人的气质与品位。隐藏在主卧内，紧临大窗又极私密的主卫和浴缸，也为空间增添了一份小小的浪漫。次卧内爽朗的蓝色搭配迷人的深咖色，干净清爽，极致且舒适的大床，搭配全软包的墙，这些都让所有来宾倍感家的温馨。

BLOOMING SUMMER FLOWERS
夏花似锦

Black and white marble is the salute and inheritance of classic British culture; the formation of modern new meaning is the perfect present of fashion and classics.

With a cup of tea, a pot of British afternoon tea and classic emerald green leather chairs, you can taste the comfort of life in a warm afternoon. Metal furnishings on the mirror make the whole space stylish. Leisure booth design extends to the edge to become a shelf. The dining room virtually becomes a reading place; with a cup of coffee and a book, you can enjoy the leisure time. The whole space is like a picturesque life scroll, which is charming, fragrant, amiable and warm. What blooms in the bedside is the most classic country rose pattern in England, which is gorgeous, full and fashionable. Living in such a place like a palace is luxurious and elegant. After a busy day, you can unload the blundering mood and be surrounded by a warm sweet and romance. Space dream may have been every boy's dream. Blue and white children's room is a dream place for the children; sleeping in the light blue bed seems to be flying when eyes are closed. As long as you have a dream, it will come true finally.

POSTMODERNISM 后现代

Design company: Ease Decoration Design
Designer: Xingbin Yang
Main materials: marble, copper, cloth, etc.
Location: Shenyang, Liaoning
Area: 103 m²

设计公司：一然设计
设计师：杨星滨
主要材料：大理石、铜、布艺等

项目地点：辽宁沈阳
项目面积：103 m²

黑白相间的大理石，是对古典英伦文化的致敬和传承，现代新意的构成方式，是时尚与古典相结合的完美呈现。

一杯红酒，一壶英式下午茶，经典翡翠绿的皮质吧椅，暖暖的午后，体味生活的惬意。镜面上金属的点缀，让整个空间不失风格。休闲卡座设计，延展到边变化为书架，餐厅无形中兼具阅读室功能，一杯咖啡，一本书，尽享悠闲时光。整个空间如同一幅如锦似画的生活画卷，馥郁芬芳且亲和温馨。床头盛开的是英国最经典的乡村蔷薇花科图案，绚丽饱满之余不失时尚。生活在像宫殿一样的空间，奢华而不失雅致。忙碌的工作之后，卸下浮躁的心情，被暖暖的温馨和浪漫包围。航天梦可能是每个男孩子都有过的理想。蓝白为主调的儿童房像是孩子梦想的空间，睡在淡蓝色的床上仿若闭上双眼就在飞翔。只要心有梦想，终会实现。

FACING THE SEA WITH SPRING BLOSSOMS

面朝大海 春暖花开

The owner has a dream about French window in mind, which is about her life tastes and feelings. As sung in the song *I wish*: "I wish I can have a bright French window, can bask in the sunshine every day, put my things on the floor, sing the past time..." This time, the designer realizes her imagination. No matter day or night, sunshine or moonlight passes through the large pieces of French window to fall on the heart; it is beautiful, warm and comfortable. In the afternoon, she can hold the cute dog, lie in the rocking chair, wait for his return from work and enjoy the beautiful scenery outside the window together.

Project name: Jingjiang City Garden
Design company: KASON
Main materials: wallpaper, parapet wall, stone, production board, etc.

Designer: Junhua Jiang
Photographer: Feng Lin

Location: Hangzhou, Zhejiang
Area: 250m²

项目名称：景江城市花园
设计公司：嘉兴康盛装饰工程有限公司
主要材料：墙纸、护墙、石材、成品板等

设计师：蒋军华
摄影师：林峰

项目地点：浙江杭州
项目面积：250m²

屋主心中有一个关于落地窗的梦想，这关乎她对生活需要的品位和感觉。正如《我希望》这首歌中唱的："我希望能拥有个明亮的落地窗，每天都能够去晒一晒太阳，把我的东西都摆在地上，再唱起从前的时光……"这次，设计师将她的想象变成了现实。无论白天或者夜晚，日光或者月光，透过大大的落地窗，正好落入心扉，感觉如此美好，很温暖，很舒服。又或者午后，抱着可爱的狗狗睡在摇椅上，等着对方下班，然后一起欣赏窗外的美景。

图书在版编目（CIP）数据

炫：引领现代居住空间新趋势 / 深圳视界文化传播有限公司编. -- 北京：中国林业出版社，2017.4
ISBN 978-7-5038-8946-2

Ⅰ．①炫… Ⅱ．①深… Ⅲ．①住宅－室内装饰设计－作品集－中国－现代 Ⅳ．① TU241

中国版本图书馆CIP数据核字（2017）第064892号

编委会成员名单
策划制作：深圳视界文化传播有限公司（www.dvip-sz.com）
总 策 划：万绍东
编　　辑：杨珍琼
装帧设计：黄爱莹
联系电话：0755-82834960

中国林业出版社 · 建筑分社
策　　划：纪　亮
责任编辑：纪　亮　王思源

出版：中国林业出版社
（100009 北京西城区德内大街刘海胡同 7 号）
http://lycb.forestry.gov.cn/
电话：（010）8314 3518
发行：中国林业出版社
印刷：深圳市雅仕达印务有限公司
版次：2017 年 5 月第 1 版
印次：2017 年 5 月第 1 次
开本：235mm×335mm，1/16
印张：20
字数：300 千字
定价：428.00 元（USD 86.00）